Walch Hands-on Science Series

Electricity and Magnetism

by Joel Beller and Kim Magliore

illustrated by Lloyd Birmingham

Project editors: Joel Beller and Carl Raab

J. WESTON

WALCH

PUBLISHER

Portland, Maine

User's Guide
to
Walch Reproducible Books

As part of our general effort to provide educational materials that are as practical and economical as possible, we have designated this publication a "reproducible book." The designation means that purchase of the book includes purchase of the right to limited reproduction of all pages on which this symbol appears:

Here is the basic Walch policy: We grant to individual purchasers of this book the right to make sufficient copies of reproducible pages for use by all students of a single teacher. This permission is limited to a single teacher and does not apply to entire schools or school systems, so institutions purchasing the book should pass the permission on to a single teacher. Copying of the book or its parts for resale is prohibited.

Any questions regarding this policy or requests to purchase further reproduction rights should be addressed to:

Permissions Editor
J. Weston Walch, Publisher
321 Valley Street • P.O. Box 658
Portland, Maine 04104-0658

1   2   3   4   5   6   7   8   9   10
ISBN 0-8251-3933-3

Copyright © 2000
J. Weston Walch, Publisher
P.O. Box 658 • Portland, Maine 04104-0658
www.walch.com

Printed in the United States of America

Contents

 # To the Teacher

This is one in a series of hands-on science activities for middle school and early high school students. A recent survey of middle school students conducted by the National Science Foundation (NSF) found that

- more than half listed science as their favorite subject.
- more than half wanted more hands-on activities.
- 90 percent stated that the best way for them to learn science was to do experiments themselves.

The books in this series seek to capitalize on these findings. These books are not texts but supplements. They offer hands-on, fun activities that will turn some students on to science. Most of these activities can be done in school, and some of them can be done at home. The authors are teachers who have field-tested the activities in a public middle school and/or high school.

This book includes activities related to a force of nature, namely, electromagnetism. Every effort has been made to use readily available, inexpensive equipment. Activities range from the simple (playing with magnets and magnetic equipment) to the complex (creating electricity by means of thermocouples and working with photovoltaic cells). There is something for every student. We strongly recommend that you try these activities yourself before asking your students to perform them.

Due to the rapid and constant evolution of the Internet, some sites may no longer be accessible at the addresses listed at the time of this printing.

THE ACTIVITIES CAN BE USED:

- to provide hands-on experiences pertaining to textbook content.
- to give verbally limited children a chance to succeed and gain extra credit.
- as the basis for class or school science fair projects or for other science competitions.
- to involve students in science club activities.
- as homework assignments.
- to involve parents in their children's science education and experiences.
- to foster an appreciation for science.

This book provides hands-on activities in which students:

- manipulate equipment.
- interpret data.
- evaluate experimental designs.
- draw inferences and conclusions.
- make predictions.
- apply the methods of science.

Each activity has a teacher resource section that includes, besides helpful hints and suggestions, a scoring rubric, a quiz, and Internet connections for those students who wish to carry out the Follow-up Activities. Instructional objectives and the National Science Standards that apply to each activity are provided in order for you to meet state and local expectations.

Generating Static Electricity

✔ INSTRUCTIONAL OBJECTIVES

Students will be able to

- construct an electroscope.
- explain how an electroscope works.
- create a static charge in a variety of ways.
- draw conclusions based on observations.

🌐 NATIONAL SCIENCE STANDARDS ADDRESSED

Students demonstrate a conceptual understanding of

- electrical motions or transformations of energy.
- big ideas and unifying concepts such as cause and effect.
- motion and forces such as net forces and magnetism.

Students demonstrate scientific inquiry and problem-solving skills by

- identifying the problem and evaluating the outcomes of the investigation.
- working individually and in teams to collect and share information and ideas.
- recognizing, analyzing, and critiquing alternative explanations.
- identifying and controlling experimental variables.

Students demonstrate effective scientific communication by

- arguing from evidence such as data collected through their own experimentation.
- explaining scientific concepts to other students.

Students demonstrate competence with the tools and technologies of science by

- using tools and technology to collect data.

✂ MATERIALS

- Plastic or rubber comb
- Piece of wool
- Pith ball
- Support stand
- Clamp holder
- Plastic ruler
- Confetti
- Two books of the same thickness (approximately 1–2 cm thick)
- Clear plastic or glass pane measuring approximately 20 cm square
- Glass stirring rod
- Piece of silk
- Small glass jar
- Sheet of aluminum foil
- Cardboard disk cut to fit the neck of the jar
- Thick (10- or 12-gauge) metal wire, 15–20 cm in length
- Tape (masking, electrical, or transparent)
- Scissors
- Glue or quick-drying cement
- Thread

HELPFUL HINTS AND DISCUSSION

Time frame: One class period
Structure: Individually or in cooperative learning groups of three students
Location: In class

In this activity, students will create static electricity and build an electroscope. Stress the fact that their hands and all materials they work with should be dry. Working with static electricity is easier when the humidity is low and the air is dry. You might want to have a standby activity on hand in case the day scheduled for this activity turns out to be humid or rainy. Our driest days occur during the winter months.

The answer to the fifth question in the Data Collection and Analysis section is that the confetti will be attracted to the negatively charged pane of glass or plastic. When the confetti hits the pane, it gains electrons and becomes negatively charged. Since like charges repel, the confetti will fall back to the table and lose its negative charge. Then the process starts over again.

<table>
<tr><td>

ADAPTATIONS FOR HIGH AND LOW ACHIEVERS

High Achievers: Encourage these students to carry out the Follow-up Activities. They should work with and assist the low achievers in the cooperative learning groups.

Low Achievers: Organize these students in cooperative learning groups with the high achievers. Provide a glossary and reference material for boldfaced terms.

</td>
<td>

SCORING RUBRIC

Full credit should be given to those students whose predictions, rationales, observations, and answers are complete and accurate. Students should answer the Concluding Questions in complete sentences. Give extra credit to those students who complete the Follow-up Activities.

</td></tr>
</table>

INTERNET TIE-INS   http://www.eskimo.com/~billb/miscon/curstat.html
http://www.mos.org/sln/toe/simpleelectroscope.html
http://www.school-for-champions.com/science/staticexpl.html

QUIZ   1. How does an electroscope store static electrical charges?
2. State the role of the following in creating a static electrical charge:
   a. rubbing
   b. low humidity
   c. nonconductors
   d. surface of a material

Generating Static Electricity

⚡ BEFORE YOU BEGIN ⚡

Static electricity is caused by the gain of electrical charges on the *surface* of a **nonconductor**. Keep in mind the word *surface* and the word *nonconductor*. They are important, as you will see when you carry out the static electricity activities in this book.

The charge is due to the transfer of **electrons** from the **atoms** of one object to another by *rubbing*. Electrons are negatively charged particles. Atoms also contain positively charged particles that are equal in number to the electrons. Thus, atoms are usually electrically neutral. Because of their position in the atom and their mass, only the electrons can be moved by rubbing. If you rub two objects together, the **friction** will loosen some of the electrons on the surface of one of the objects. The greater the number of electrical charges on the surface of the nonconductor, the stronger the charge will be. When an object loses electrons from the atoms on its surface, that object will become **positively charged** (+). The object that gains those electrons will become **negatively charged** (–). If an object is a nonconductor the charges will remain on its surface. Rubber, wood, and fabrics are nonconductors with surfaces. Air is a nonconductor, but since it is a gas, it has no surface on which the electrons can collect.

Conductors, which include the metals, such as iron or aluminum, act differently. The electrons easily flow to the interior of these materials. Thus, conductors will neutralize both negatively and positively charged nonconductors.

In this activity you will create static electricity and build an **electroscope**, which is a sensitive instrument for detecting the presence of static electricity. Electroscopes can also store static electricity for varying periods of time. An electroscope consists of a metal rod, with two moveable metal leaves enclosed in a glass container as shown in the diagram in the Procedure section ("The Electroscope").

✂ MATERIALS

- Plastic or rubber comb
- Piece of wool
- Pith ball
- Support stand
- Clamp holder
- Plastic ruler
- Confetti
- Two books of the same thickness (approximately 1–2 cm thick)
- Clear plastic or glass pane measuring approximately 20 cm square

- Glass stirring rod
- Piece of silk
- Small glass jar
- Sheet of aluminum foil
- Cardboard disk cut to fit the neck of the jar
- Thick (10- or 12-gauge) metal wire, 15–20 cm in length
- Tape (masking, electrical, or transparent)
- Scissors
- Glue or quick-drying cement
- Thread

(continued)

3

Generating Static Electricity *(continued)*

PROCEDURE

Write all your answers to the questions in the Data Collection and Analysis section.

1. Attach a piece of sewing thread approximately 20 cm in length to the pith ball with glue or quick-drying cement.

2. After the glue has dried, suspend the ball from the clamp holder by the thread.

3. Rub the comb with the piece of wool several times, as though cleaning it. Then bring the comb close to the pith ball. **Don't touch the pith ball with the comb just yet!** Do this several times. What do you observe? What is the explanation for your observation?

4. Now, touch the pith ball with the comb. What do you observe? What is the explanation for your observation?

5. Bring your index finger close to the charged pith ball. What do you observe? What is the explanation for your observation?

6. Place the two books about 15 cm apart on the tabletop.

7. Sprinkle some confetti on the table between the two books. Support the glass or plastic pane on the books, covering the confetti.

8. Rub the top surface of the glass or plastic with the wool briskly but gently.

9. Observe the confetti for 3–4 minutes. What does it do? How can you explain the action?

10. Remove the glass, run the plastic comb through your hair several times and then bring it close to the confetti. How do your observations compare with those made in Step 9?

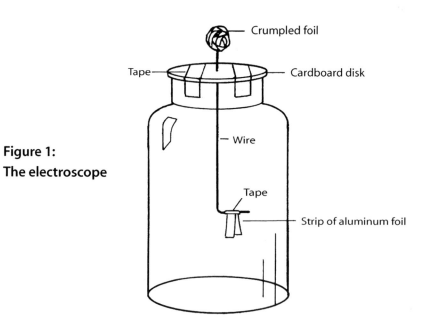

**Figure 1:
The electroscope**

(continued)

Generating Static Electricity (continued)

11. Using the wire as a tool, make a hole in the center of the cardboard disk. Push the wire through the hole.

12. Bend one end of the wire 90 degrees, as shown in the diagram above.

13. Cut a strip of aluminum foil about 5 cm long and 1 cm thick.

14. Fold the strip in half and then tape it over the bent end of the wire to create two flaps.

15. Place the bent end of the wire into the jar and tape the cardboard disk onto the neck of the jar. Be sure 3 or 4 cm of wire protrudes through the top of the disk.

16. Crumple the piece of foil into a ball and place it on top of the wire sticking out of the disk. (Figure 1)

17. Run a plastic or rubber comb through your hair several times to give it a charge.

18. Touch the crumpled foil ball.

19. Repeat Steps 17 and 18 several times. Observe what happens to the flaps of aluminum when the comb touches the metal.

DATA COLLECTION AND ANALYSIS

1. What did you observe when the comb was brought close to the pith ball? How do you explain this observation? _____

_____

2. What happened when you touched the pith ball with the comb? Why did this happen? _____

_____

3. What happened when you brought your finger close to the charged pith ball? Why did this happen? _____

_____

4. What happened when you touched the pith ball with your finger? Explain the cause.

_____

_____

5. What happened to the pieces of confetti under the pane? How can you explain their action?

_____

_____

6. How did the observations you made while using your comb compare with those you made using the pane and the wool? How do you explain your observations?_____

_____

_____

(continued)

Generating Static Electricity *(continued)*

7. What happened to the aluminum flaps in the electroscope when you touched the crumpled ball with the charged comb? _____

_____

 CONCLUDING QUESTIONS

1. What is the relationship between friction and static electricity?_____

_____

_____

_____

_____

2. How can "static cling," that is, pieces of clothing sticking to each other when they are removed from a clothes dryer, be explained by what you have learned in this activity? _____

_____

_____

_____

_____

⚡ Follow-up Activities ⚡

1. Take two balloons that have been inflated to the same diameter and tie a string around the neck of each. Hold the two strings in one hand, allowing the balloons to dangle freely, and record what happens to the balloons. Rub each balloon with a wool cloth and let them once again hang freely from the strings. What happened to the balloons? Explain why this occurs.

2. Research how Whimhurst and Van de Graaff generators create static electricity. Report your findings to your teacher. After obtaining your teacher's permission, demonstrate either generator to your classmates.

3. Design an experiment to see if the static charges are more concentrated at the tip of a comb or glass rod rather than in the central part. Present your experimental plan to your teacher and then carry out the experiment. Report your findings to your classmates.

4. Investigate "antistatic" products, including solids such as Bounce™ as well as sprays. Write a report on how they work.

Tiny Sparks and Big Sparks (Lightning)

✔️ INSTRUCTIONAL OBJECTIVES

Students will be able to

- create static electrical sparks.
- name fabrics that are good spark producers.
- describe two ways to reduce the sparking effect of static electricity.
- draw conclusions based on observations.

🌐 NATIONAL SCIENCE STANDARDS ADDRESSED

Students demonstrate a conceptual understanding of

- electrical motions or transformations of energy.
- big ideas and unifying concepts such as cause and effect.
- motion and forces such as net forces and magnetism.

Students demonstrate scientific inquiry and problem-solving skills by

- identifying a problem and evaluating the outcomes of the investigation.
- working individually to collect and share information and ideas.
- recognizing, analyzing, and critiquing alternative explanations.
- identifying and controlling experimental variables.

Students demonstrate effective scientific communication by

- arguing from evidence such as data collected through their own experimentation.
- explaining scientific concepts to other students.

Students demonstrate competence with the tools and technologies of science by

- using tools and technology to collect data.

✂️ MATERIALS

- Sneakers
- Shoes with leather soles
- Metal key or nail
- Rug (wool and/or synthetic fibers)
- Wool socks
- Cotton socks
- Silk/nylon stockings
- Synthetic fabric socks
- Blouses and sweaters made of various fabrics

HELPFUL HINTS AND DISCUSSION

Time frame: One class period
Structure: Individually
Location: At home

In this activity, your students will create sparks by shuffling across a rug wearing different pieces of clothing, then touching a metal doorknob. Then, they will determine what kinds of fabrics create sparks when they change clothing or shoes and socks.

Give your students guidance regarding terms to use to describe the sparks, since this activity will be done at home on an individual basis. Such terms might be "large," "small," "annoying," etc. It would be wise to tabulate the results of the entire class in the form of a chalkboard table. Then, a comparison could be made between the types of rug material and the sparking effect produced.

Provide a teacher demonstration of lightning using a Wimshurst generator or Van de Graaff generator. Caution the students against touching high voltage sparks.

ADAPTATIONS FOR HIGH AND LOW ACHIEVERS

High Achievers: Encourage these students to carry out the Follow-up Activities.

Low Achievers: Provide a "walk-through" demonstration of how this activity should be done, since students will do it at home. Provide some fictional data as a sample. Invite the students to have their parents help them carry out this activity.

SCORING RUBRIC

Full credit should be given to those students whose observations and answers are complete and accurate. Students should answer the Concluding Questions in complete sentences. Give extra credit to those students who complete the Follow-up Activities.

**INTERNET TIE-INS**   http://www.nassauredcross.org/sumstorm/thunder1.htm
http://www.school-for-champions.com/science/staticspark.htm
http://www.school-for-champions.com/science/staticgen.htm

QUIZ   1. State two ways in which static electric sparks can be reduced.
2. State the role of the following in creating a static electrical spark:
   a. dry skin
   b. type of fabric
3. Name a fabric that does NOT tend to build up a static electricity charge.

Tiny Sparks and Big Sparks (Lightning)

〰 BEFORE YOU BEGIN 〰

✋ **WARNING: This activity should not be done on a humid summer day. The best results will be achieved if the humidity is low. Winter days are usually much drier than summer days. This activity can be done in humid weather in a room that has had its moisture removed by a dehumidifier!**

You are already familiar with static electricity attracting or repelling nonconductors. Static electricity can also cause sparks. Lightning is really a very large spark. A spark produced when you walk across a rug and touch a metal light switch may be 6 millimeters in length, while a bolt of lightning can be 100 kilometers long!

In the previous activity, you learned that charges collect on the surface of nonconductors. However, it is extremely doubtful that you created a spark, no matter how vigorously you rubbed. In order to create sparks, you need conductors. Metals are very good conductors. People also conduct electricity, but not nearly as effectively as metals. The salts in your blood and other body fluids are responsible for making you a conductor. Have you ever walked across a rug and touched someone and a spark occurred? If you kissed someone on the lips, the spark would be annoying, but not strong enough to be dangerous.

Technically, the spark is due to the rapid jumping of many electrons from one conductor to another through the air. The electrons bump into the air molecules at high speed. This heats up the air. The air glows white-hot. This is a spark that you can see and often feel. Once the electrons have leaped to the other conductor, the air quickly cools down and the sparking event is over.

A lightning bolt is just a large spark. How the huge sparks we call lightning occur is not really perfectly clear to scientists. They do know that tall thunderclouds are generally negative at their **bases**, while the upper parts are positive. The most familiar type of lightning bolts are from the base of the cloud to the ground. These, of course, are the most dangerous. The lightning will follow the shortest path between the base of the cloud and the ground. If you are standing in a meadow or on a lawn during a lightning storm, you may be the tallest object on the ground. If this is the case, the lightning bolt will strike you on its way to the ground. Other lightning bolts flash between the positive and negative portions of a cloud. These lightning bolts are not dangerous.

In this activity you will investigate sparks caused by static electricity and learn ways of avoiding sparks and shocks.

✂ MATERIALS

- Sneakers
- Shoes with leather soles
- Metal key or nail
- Rug (wool and/or synthetic fibers)
- Wool socks

- Cotton socks
- Silk/nylon stockings
- Synthetic fabric socks
- Blouses and sweaters made of various fabrics

(continued)

Tiny Sparks and Big Sparks (Lightning) *(continued)*

PROCEDURE

To make your findings valid, carry out all procedural steps during the same day on the same rug.

1. Work in a partially darkened room whose floor is covered by a rug. Indicate in the Data Collection and Analysis section the type of material the rug is made of. Stand on the end of the rug, facing a wall in which there is a door with a metal doorknob. If the room has an area rug, stand on the far edge. If the floor is covered with wall-to-wall carpeting, stand against the wall opposite the door.

2. While wearing sneakers or shoes with rubber soles, shuffle your feet and go to the metal doorknob. Touch it. Did you see a spark or feel any discomfort? Record your data for this step and all other steps in the tables in the Data Collection and Analysis section.

3. Repeat Step 2, but this time wear shoes with leather soles.

4. Repeat Step 2, this time wearing cotton socks and no shoes.

5. Repeat Step 2, this time wearing silk or nylon stockings and no shoes.

6. Repeat Step 2, this time wearing socks that are made of different synthetic fibers or a blend of synthetic fibers or cotton or wool, and no shoes.

7. Repeat Step 2, but this time shuffle across the rug in bare feet.

8. Repeat the step in which you saw the strongest spark, except this time put a metal key or iron nail in your hand and touch the doorknob with the key. Compare the spark and the sensation with those seen and felt the first time. Record your description in the Data Collection and Analysis section.

9. Repeat another step that produced a spark and again touch the doorknob with the metal object.

10. Carry out the next phase of this activity after dark in your room. Try on and take off a variety of sweaters, blouses, shirts, etc., made of different fabrics, including cotton, wool, silk, linen, and synthetics, such as polyesters. Which ones cause sparks when you remove them? Record the results of all of these in the Data Collection and Analysis section.

11. Repeat Step 10, but ground yourself first by touching the doorknob, then removing the garment. Compare your data with that obtained in Step 10.

(continued)

Name _____ Date _____

Tiny Sparks and Big Sparks (Lightning) (continued)

DATA COLLECTION AND ANALYSIS

Type of rug: _____

TABLE 1

Foot Covering	Spark (Describe)	Annoying Sensation (Yes/No)
Rubber-soled shoes		
Leather-soled shoes		
Cotton socks		
Silk/nylon stockings		
Synthetic fiber socks (Fabric: _____)		
Bare feet		
Step ___ + metal		
Step ___ + metal		

TABLE 2

Item of Clothing—Type of Fabric (Without Grounding)	Spark (Describe)	Spark after Grounding (Describe)

(continued)

Tiny Sparks and Big Sparks (Lightning) *(continued)*

❓ CONCLUDING QUESTIONS

1. Which type of footwear produced the largest spark? _____

2. Which type of footwear produced the least spark or no spark? _____

3. What was the effect of touching the doorknob with the metal object? _____
   _____

4. Which fabric rubbed on your skin produced the most sparking? _____

5. Which fabric produced the least or no sparking? _____

6. What was the effect of grounding upon the sparks produced by rubbing your skin with various
   fabrics? _____
   _____

⚡ Follow-up Activities ⚡

1. Repeat Step 10, but first apply a moisturizer to your skin. How do these results compare to
   those you got at first?

2. Repeat Steps 1–6, but first run a humidifier in your room for a half hour. Again, compare
   these results with those you got previously.

3. Determine if your pajamas and bed sheets cause sparking when you get into bed. Note what
   materials the pajamas and the sheets are made of. Report your results to the class.

Static Eliminators: Is the Internet Always Right?

☑ INSTRUCTIONAL OBJECTIVES

Students will be able to

- design experiments to test hypotheses.
- draw conclusions based on observations.
- compare the efficacy of several antistatic products.

🌐 NATIONAL SCIENCE STANDARDS ADDRESSED

Students demonstrate a conceptual understanding of

- electrical motions or transformations of energy.
- big ideas and unifying concepts such as cause and effect.

Students demonstrate scientific inquiry and problem-solving skills by

- identifying the problem and evaluating the outcomes of the investigation.
- working individually and in teams to collect and share information and ideas.
- recognizing, analyzing, and critiquing alternative explanations.

Students demonstrate effective scientific communication by

- arguing from evidence such as data collected through their own experimentation.
- explaining scientific concepts to other students.

✂ MATERIALS

- Packages of two or three different brands of antistatic sheets for use in a clothes dryer
- Varied—depending upon claims for antistatic sheets investigated
- Sealable plastic bags
- Marking pens

HELPFUL HINTS AND DISCUSSION

Time frame: One class period for preparation and discussion
Structure: Individual
Location: In class and at home

In this activity, students will learn of other uses for the antistatic sheets used to prevent clothes from clinging to each other in the clothes dryer (static cling) as suggested by a page on the Internet that was sent to one of the authors. Be sure to relate static cling to the topic of static electricity.

Introduce the terms **ion** and **cation**, which appear on the labels of antistatic sheets. You can purchase antistatic sheets made by various manufacturers at any supermarket. We suggest that each student test two or three different brands. Check the ingredients carefully to be sure they are the same. If not, bring this to the attention of the students. Some of the suggested uses may not be practical to test for. For example, only female students may want to test pantyhose, and mosquitoes might not live in your area at the time you want the students to carry out this activity. Divide up the suggestions that can be done easily among the class. Perhaps five suggestions could be investigated by each pupil. Overlapping would be in order to get more experimental data. Discuss with the students how the statements in the e-mail can be investigated scientifically with controls.

ADAPTATIONS FOR HIGH AND LOW ACHIEVERS

High Achievers: Have these students carry out the Follow-up Activities. They also should help the low achievers design their experiments.

Low Achievers: Provide a glossary and reference material for boldfaced terms in this activity. Students' experimental designs will need careful scrutiny. Students should be provided with detailed instructions as to how to carry out their experiments. Suggest to them that an adult assist them.

SCORING RUBRIC

Full credit should be given to those students whose observations and answers are complete and accurate. Students should answer the Concluding Questions in complete sentences. Give extra credit to those students who complete the Follow-up Activities.

INTERNET TIE-INS   http://www.phoenix-lic.com/cgi-local/shop.pl/page=index.html
http://www.leesnewsome.co.uk/household.htm

QUIZ   1. Why are antistatic sheets added to the dryer when drying clothes?
2. How do clothes build up static charges in a clothes dryer?

Static Eliminators: Is the Internet Always Right?

⚡ BEFORE YOU BEGIN ⚡

You can find many interesting statements on the World Wide Web. Web pages deal with many subjects ranging from science to purchasing an automobile. Is everything on the Internet the truth? In this activity you will have a chance to find out.

Below is a message that came from the Web. The name of the product has been replaced by an "X." The product's main job is to prevent **static cling**, which occurs when clean, dry clothes are removed from a clothes dryer. You are probably familiar with static cling, which is really the attraction of one fabric for another because of static electricity. If you have ever removed clothing from a dryer and had to pull the garments apart and gotten small shocks in the process, you know what static cling is. Product X also claims to make the clothing that is dried softer. And it claims to freshen and remove musty odors from drawers and suitcases. There are a number of products similar to X made by different companies. Most have the same basic ingredients. That is, **biodegradable cationic** softeners and perfumes. Product X has both!

The e-mail message below states that Product X can do many other helpful things, from repelling mosquitoes to cleaning TV screens. Look over the message below and decide which claims for Product X you wish to and are able to research. Then, perhaps you can think of other uses for Product X that your teacher will allow you to investigate.

Subj: Product X
Date: 9/18/99 7:39:19 A.M. Eastern Daylight Time
From: CZ
To: Author

Product X—the stuff you use in your dryer

- Repels mosquitoes. Tie a sheet of X through a belt loop when outdoors during mosquito season.
- Eliminates static electricity from your television screen. Since X is designed to help eliminate static cling, wipe your television screen with a used sheet of X to keep dust from resettling.
- Dissolves soap scum from shower doors. Clean with a used sheet of X.
- Freshens the air in your home. Place an individual sheet of X in a drawer or hang one in the closet.
- Prevents thread from tangling. Run a threaded needle through a sheet of X to eliminate the static cling on the thread before sewing.
- Eliminates static cling from pantyhose. Rub a damp, used sheet of X over the hose.
- Prevents musty suitcases. Place an individual sheet of X inside empty luggage before storing.
- Freshens the air in your car. Place a sheet of X under the front seat.
- Cleans baked-on food from a cooking pan. Put a sheet in the pan, fill it with water, let it sit overnight, and sponge clean. The antistatic agents apparently weaken the bond between the food and the pan, while the fabric softening agents soften the baked-on food.
- Eliminates odors in wastebaskets. Place a sheet of X at the bottom of the wastebasket.
- Collects cat hair. Rubbing the area with a sheet of X will magnetically attract all the loose hairs.
- Eliminates static electricity from venetian blinds. Wipe the blinds with a sheet of X to prevent dust from resettling.
- Wipes up sawdust from drilling or sandpapering. A used sheet of X will collect sawdust like a tack cloth.
- Eliminates odors in dirty laundry. Place an individual sheet of X at the bottom of a laundry bag or hamper.

(continued)

Static Eliminators: Is the Internet Always Right? *(continued)*

MATERIALS

- Packages of two or three different brands of antistatic sheets for use in a clothes dryer
- Varied—depending upon claims for antistatic sheets investigated

- Sealable plastic bags
- Marking pens

PROCEDURE

1. Decide which statements in the e-mail you wish to investigate. Get your teacher's approval.

2. Your teacher will give you two or three sheets, each a different brand. Place the antistatic sheets in individual plastic bags. Seal the bags in order to keep the perfume and other chemicals from escaping into the air. Label each bag with the brand name so you know which sheet is in each bag.

3. Write an experimental plan for researching a minimum of 5 of the 15 statements in the e-mail. In your plan, be sure to include a method for comparing each of the brands of antistatic sheets. Don't forget to include a control for each statement investigated.

4. Include any additional ideas you may have for using antistatic sheets.

5. After your teacher has approved your plan, carry it out at home.

6. After experimenting, enter your data in the Data Collection and Analysis section.

7. On the assigned day, bring your data to class for discussion and comparison with the results of other students.

DATA COLLECTION AND ANALYSIS

Statement Tested	Product	Result

(continued)

Static Eliminators: Is the Internet Always Right? *(continued)*

1. Which statements that you experimented with proved to be true? _____

    _____

    _____

2. Which did you prove were false? _____

    _____

    _____

3. Which product worked the best for each statement, or did all the products tested work the same?

    _____

    _____

4. Which product, if any, would you consider to be the best? Explain your answer. _____

    _____

    _____

❓ CONCLUDING QUESTIONS

1. Why are perfumes added to antistatic products? _____

    _____

2. How do you think these products remove "static cling"? _____

    _____

⚡ Follow-up Activities ⚡

1. Repeat your experiments using an antistatic spray, which can be bought in the laundry-soap section of a local store. How do your results compare?

2. Research cations and their use in water purification. Report your findings to the class.

Playing with Cereal, Marbles, and Other Magnetic Materials

✔ INSTRUCTIONAL OBJECTIVES

Students will be able to

- determine the north and south poles of various magnets.
- draw conclusions based on observations.
- determine what magnetic substances have in common.
- test cereals for their relative iron content.

🌐 NATIONAL SCIENCE STANDARDS ADDRESSED

Students demonstrate a conceptual understanding of

- big ideas and unifying concepts such as cause and effect.
- motion and forces such as net forces and magnetism.

Students demonstrate scientific inquiry and problem-solving skills by

- evaluating the outcomes of the investigation.
- working in teams to collect and share information and ideas.
- recognizing, analyzing, and critiquing alternative explanations.
- identifying and controlling experimental variables.

Students demonstrate effective scientific communication by

- arguing from evidence such as data collected through their own experimentation.

Students demonstrate competence with the tools and technologies of science by

- using tools and technology to collect data.

MATERIALS

- Magnetic rubber strip, 10 cm in length
- Scissors
- Two ceramic ring magnets
- Drinking straw
- Two disk magnets
- Two horseshoe magnets
- Two steel bar magnets with poles marked
- Alnico bar magnet
- Breakfast cereal that is rich in iron (100% of Daily Value)
- Breakfast cereal that is poor in iron (25% or less of Daily Value)
- Soda can
- Vegetable can
- Pieces of glass, paper, plastic, wood, rubber, etc.
- Cobalt magnet
- Samples of various metals (copper, zinc, brass, soft iron, steel, etc.)
- Two magnetic marbles
- Support stand and clamp holder
- String
- Marking pen
- Box of metal paper clips

HELPFUL HINTS AND DISCUSSION

Time frame: One to two class periods
Structure: Cooperative learning groups of three students
Location: In class

In this activity, students will learn which substances are magnetic, and that magnets can be made in many different shapes and from different materials. Explain what the term "Alnico" means. Demonstrate to the class how to crush the cereals in one hand using the thumb of the other hand as the pestle.

ADAPTATIONS FOR HIGH AND LOW ACHIEVERS

High Achievers: Stress the need for as much accuracy as possible in collecting the data in this activity. Encourage these students to carry out the Follow-up Activities. They should also work with and assist the low achievers in the cooperative learning groups.

Low Achievers: Have the high achievers assist these students in creating the "confirmation setup." Encourage these students to serve as experimenters in this activity.

SCORING RUBRIC

Full credit should be given to those students whose observations and answers are complete and accurate. Students should answer the Concluding Questions in complete sentences. Give extra credit to those students who complete the Follow-up Activities.

INTERNET TIE-INS   http://science.msfc.nasa.gov/ssl/pad/sppb/edu/
http://www.technicoil.com/magnetism.html

QUIZ   1. Name three materials that are used to make magnets.
2. Why are some magnets made of steel rather than soft iron?
3. Name two metals that are not magnetic.

Name _____ Date _____

Playing with Cereal, Marbles, and Other Magnetic Materials

⚡ BEFORE YOU BEGIN ⚡

There are three parts to this activity, which deals with magnets and the substances that are attracted to them. You already know some things about magnets. For example, bar magnets made from iron or steel have a north and south pole, and like poles **repel** each other while unlike poles **attract** each other. Now you will determine if this attraction and repelling is true for magnets made of other materials and having different shapes. You will experiment with ceramic magnets, magnetic marbles, and ring magnets. You will also investigate whether other materials besides iron and steel can be made into magnets. Finally, your team will determine which common items, such as the cold cereal you eat for breakfast, are magnetic, and which are not.

✂ MATERIALS

- Magnetic rubber strip, 10 cm in length
- Scissors
- Two ceramic ring magnets
- Drinking straw
- Two disk magnets
- Two horseshoe magnets
- Two steel bar magnets with poles marked
- Alnico bar magnet
- Breakfast cereal that is rich in iron (100% of Daily Value)
- Breakfast cereal that is poor in iron (25% or less of Daily Value)

- Soda can
- Vegetable can
- Pieces of glass, paper, plastic, wood, rubber, etc.
- Cobalt magnet
- Samples of various metals (copper, zinc, brass, soft iron, steel, etc.)
- Support stand and clamp holder
- String
- Two magnetic marbles
- Marking pen
- Box of metal paper clips

📋 PROCEDURE

1. Set up the support stand and clamp as shown in the diagram below. Suspend a steel bar magnet, whose poles have been indicated, from the clamp using the string. Center the string around the magnet and tie the string in place so that the magnet will swing freely, as shown in Figure 1.

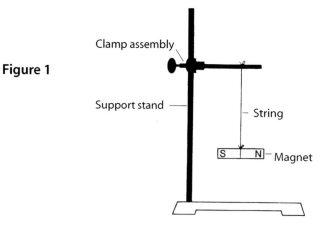

Figure 1

Clamp assembly

Support stand

String

S | N — Magnet

(continued)

Playing with Cereal, Marbles, and Other Magnetic Materials *(continued)*

2. Using the other steel bar magnet, confirm that like poles repel each other and unlike poles attract each other. This "confirmation setup" will be used to determine the north and south poles of the various magnets you will investigate.

3. Using the confirmation setup and a magnetic marble, determine if a magnetic marble has a north and south pole. Record your answer in the Data Collection and Analysis section. If you do locate a north and south pole on the marble, mark their positions with the marking pen. Use the symbols "N" and "S" for north and south poles.

4. After you have located the poles on one marble, do the same for the second magnetic marble.

5. Using both marbles, determine if the N and S poles attract while the S-S poles and N-N poles repel. Record your answer in the Data Collection and Analysis section.

6. Repeat Step 3, but replace the magnetic marble with a ceramic ring magnet. The rings are made of metallic oxide powders that are pressed together under high pressure.

7. Slip the marked ceramic ring magnet over a drinking straw that a member of your group is holding upright on the table top. Slip the other ceramic ring magnet over the straw and let it fall toward the marked ring magnet. What do you observe? What is the explanation for this observation? Record your observation and explanation in the Data Collection and Analysis section.

8. Remove the unmarked magnet from the straw, turn it upside down, and again slip it over the straw, as you did in Step 7. What do you observe this time? What is the explanation for this observation? Again, record your observation and explanation in the Data Collection and Analysis section.

9. Using appropriate techniques, determine if the two disk magnets, which are also ceramic, have north and south poles. If so, mark their location with an "N" and an "S." Record your answer in the Data Collection and Analysis section.

10. Using the scissors, cut off two pieces of the magnetic rubber strip. Make each piece approximately 2 cm in length. The strip is really made of flexible plastic, not rubber. Magnetic grains are embedded in the plastic. Again, using appropriate techniques, determine if these two strips have a north and a south pole. If so, mark them with the marking pen and record your results in the Data Collection and Analysis section.

11. Compare the strength of the magnets (bar, horseshoe, ring, disk, etc.) at your disposal. Test their relative strengths by counting the number of paper clips they can pick up when each magnet is inserted in a box of paper clips. Record your observations in Table 1. Which magnet picked up the most paper clips?

12. You will now determine what materials are attracted to magnets. Use the confirmation setup to test the soda can, vegetable can, and the various metals, plastics, paper, wood, rubber, etc., that are available. Record your observations in Table 2 in the Data Collection and Analysis section.

(continued)

Playing with Cereal, Marbles, and Other Magnetic Materials *(continued)*

13. Have one member of your group take a few pieces of the cereal that is rich in iron content in one hand and, using the thumb of the other hand, crush the cereal pieces into very small particles. Have a second team member put one end of the Alnico bar magnet into the cereal particles, then withdraw the magnet. Do any of the particles stick to the magnet? Record your results in Table 3 in the Data Collection and Analysis section.

14. Have a different member of your group take a few pieces of the cereal that is poor in iron content in one hand and, using the thumb of the other hand, crush the cereal pieces into very small particles. Have another team member put one end of the Alnico bar magnet into the cereal particles, then withdraw the magnet. Do any of the particles stick to the magnet? Record your results in Table 3 in the Data Collection and Analysis section. Compare the amount of high-iron-content cereal that stuck to the magnet with the amount of low-iron-content cereal that stuck to the magnet.

DATA COLLECTION AND ANALYSIS

1. Does a magnetic marble have a north and a south pole? _____ If so, where are they located? _____ What is the reasoning behind your answers? _____
   _____

2. Do magnetic marbles obey the law of magnetic poles, which says "Like poles repel each other and unlike poles attract"? _____

3. Do the ceramic ring magnets have a north and a south pole? _____ If so, where are they located? _____ Do they obey the law of magnetic poles? _____ What is your evidence for your answer? _____
   _____

4. Do the ceramic disk magnets have a north and a south pole? _____ If so, where are they located? _____ Do they obey the law of magnetic poles? _____ What is your evidence for your answer? _____
   _____

5. Do the magnetic flexible strips have a north and a south pole? _____ If so, where are they located? _____ Do they obey the law of magnetic poles? _____ What is your evidence for your answer? _____
   _____

(continued)

Name _____ Date _____

Playing with Cereal, Marbles, and Other Magnetic Materials *(continued)*

TABLE 1

Type of Magnet	Number of Paper Clips Picked Up
Steel bar magnet	
Alnico bar magnet	
Horseshoe magnet	
Magnetic marble	
Ceramic ring magnet	
Ceramic disk magnet	
Cobalt magnet	
Flexible magnetic strip	

6. Which magnet picked up the most paper clips? _____

7. Which magnet picked up the fewest paper clips? _____

TABLE 2

Material	Magnetic (Yes/No)
Brass	
Soft iron	
Steel	
Zinc	
Copper	
Vegetable can (iron coated with a layer of tin)	
Soda can (aluminum)	
Plastic	
Wood	
Rubber	
Glass	

(continued)

Playing with Cereal, Marbles, and Other Magnetic Materials *(continued)*

8. Which substances were magnetic? _____
   _____

9. Which substances were not magnetic? _____
   _____

TABLE 3

Type of Cereal	Relative Amount of Cereal Picked Up
Iron "rich"	
Iron "poor"	

10. Which cereal contained the most iron? _____ How do you know? _____
    _____
    _____

❓ CONCLUDING QUESTIONS

1. Do all magnets obey the law of magnetic poles? _____ Explain your answer. _____
   _____
   _____

2. What types of substances are magnetic? _____
   _____
   _____

⚡ Follow-up Activities ⚡

1. Research in the library and on the Internet liquids with magnetic properties.

2. Investigate cow magnets. Find out the origin of their name.

3. Using three or four ring magnets and a thin wooden dowel in a wad of clay to support the dowel, float several magnets in the air, one over the other.

Finding Your Way Using a Magnet

✔️ INSTRUCTIONAL OBJECTIVES

Students will be able to

- construct a compass.
- draw conclusions based on observations.
- explain how a compass works.
- use a field compass to travel from one place to another.
- explain how a gyrocompass works.

🌐 NATIONAL SCIENCE STANDARDS ADDRESSED

Students demonstrate a conceptual understanding of

- electrical motions or transformations of energy.
- big ideas and unifying concepts such as cause and effect.
- motion and forces such as net forces and magnetism.

Students demonstrate scientific inquiry and problem-solving skills by

- evaluating the outcomes of the investigation.
- working in teams to collect and share information and ideas.
- identifying and controlling experimental variables.

Students demonstrate effective scientific communication by

- arguing from evidence such as data collected through their own experimentation.
- explaining scientific concepts to other students.

Students demonstrate competence with the tools and technologies of science by

- using tools and technology to collect data.

✂️ MATERIALS

- Saucer or plastic pie plate
- Small flat piece of cork, approximately 1–2 cm in diameter
- Transparent tape
- 🖐️ Sewing needle, approximately 3 cm in length
- Field (orienteering) compass with a direction-of-travel needle
- Bar magnet
- Water
- *Optional*—gyroscope

🖐️ = Safety icon

HELPFUL HINTS AND DISCUSSION

Time frame: One to two class periods
Structure: Cooperative learning groups of two students
Location: In class or at home

In this activity, students will construct and use a compass. Stress to students the importance of removing all ferrous materials from the table and their person when using compasses. Point out that the needle is sharp and should be handled in a safe manner. Using the compass on the baseball field could be done during the second period devoted to this activity or as an after-school activity.

ADAPTATIONS FOR HIGH AND LOW ACHIEVERS

High Achievers: Stress the need for accuracy in collecting the data when working with the field compass. Encourage these students to carry out the Follow-up Activities. They should also work with and assist the low achievers in the cooperative learning groups.

Low Achievers: Organize these students in cooperative learning groups with the high achievers. Discuss the need for cooperation among the members of each group. Provide a glossary and reference material for boldfaced terms. Assist these students in answering question 7 in the Data Collection and Analysis section and question 2 of the Concluding Questions.

SCORING RUBRIC

Full credit should be given to those students whose observations and answers are complete and accurate. Students should answer the Concluding Questions in complete sentences. Give extra credit to those students who complete the Follow-up Activities.

INTERNET TIE-INS   http://www.howstuffworks.com/compass.htm
http://www.uio.no/~kjetikj/compass/

QUIZ   1. Why is a compass with a direction-of-travel arrow useful?
2. What is meant by magnetic deviation?

Finding Your Way Using a Magnet

⚡ BEFORE YOU BEGIN ⚡

People have used compasses for hundreds of years to find their way from place to place. Compasses are still found on planes and ships, and even some cars have compasses. A compass is really a sensitive detector of weak magnetic fields. It is a magnetic needle or pointer that is supported at its center of gravity so that it can rotate freely. One end of the needle is marked in some way to indicate that it points to the north.

As you may know from a previous activity, magnets have a **north** and a **south** pole. If you bring like poles together, that is, the north poles of two magnets or the south poles of two magnets, the like poles will **repel** each other. On the other hand, two unlike poles, a north and a south pole, will **attract** each other.

A compass works because the earth's core is similar to a giant bar magnet. At the earth's surface the magnetic field is weak. However, it is strong enough to be detected by a compass needle balanced on an almost **frictionless pivot** point. Thus, a south pole, which we can also refer to as a **north-seeking pole** of a compass, will point to the north magnetic pole of the earth.

However, there is a problem! The north magnetic pole of the earth and the north geographic pole of the earth are *not* in the same place. The north magnetic pole is located in Canada. It is some *1,100 miles south* of the geographic north pole. This means that, in most cases, the compass is pointed to the magnetic north pole, not the geographic north pole. The angle between the compass needle pointing north and a line drawn between the north and south geographic poles is called the **declination**. This angle varies between 25° east and 20° west in the United States. For example, the declination for New York City is approximately 13°W. This means that a compass needle in New York City will point 13 degrees west of geographic north. The declination changes from year to year by a small amount and must be taken into account when using a compass to get from place to place. Expensive compasses can be adjusted to compensate for the declination.

✂ MATERIALS

- Saucer or plastic pie plate
- Small flat piece of cork, approximately 1–2 cm in diameter
- Transparent tape
- ✋ Sewing needle, approximately 3 cm in length
- Field (orienteering) compass with a direction-of-travel needle

- Water
- Bar magnet
- *Optional*—gyroscope

✋ = Safety icon

(continued) 🔥

Finding Your Way Using a Magnet *(continued)*

PROCEDURE

1. Work on a wooden table or one that does not have any materials on it, such as stainless steel bars or iron gas pipes, that will be attracted to a magnet. Fill the saucer or pie plate with water until its depth is approximately 2 cm. Place it on the table.

2. Stroke the sewing needle with the bar magnet about 15 times, always in the same direction, as shown in Figure 1.

3. Tape the needle to the piece of cork and gently float it in the water, as shown in Figure 2.

4. Observe what happens to the cork. Record your results in the Data Collection and Analysis section.

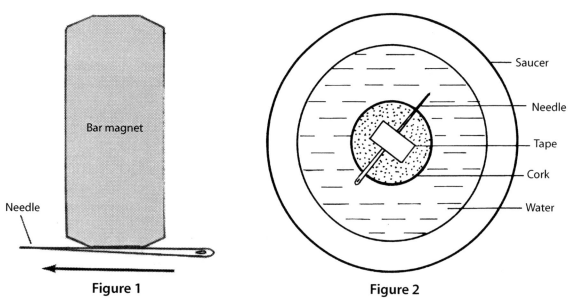

Figure 1   **Figure 2**

5. Place the field compass near your homemade compass. Be sure to remove from your pockets or the tabletop the bar magnet, your watch, and any materials, such as paper clips, that might affect the movement of both compasses. Place these things away from your work area. Why must these precautions be taken?

6. Compare the directions in which both needles are facing. How do they compare? Record your answer in the Data Collection and Analysis section.

7. Which end of your homemade compass points to the earth's north magnetic pole, the pointed end or the end with the hole? Record your answer in the Data Collection and Analysis section.

8. Walk around the room holding the field compass, starting with the needle pointing to the magnetic north. Imagine what would happen if you were traveling in a jet plane at a speed of 300 miles per hour. Why would your field magnet be unsatisfactory for navigation of a jet plane or a rapidly moving ship? Record your answer in the Data Collection and Analysis section.

(continued)

9. *Optional*—Modern planes and ships use **gyroscopic compasses** as well as a magnetic compass. When a gyroscope is spun, and if supported in a frame, it will maintain the same direction, even if the frame is moved or rotated. Using a magnetic compass as a starting point, the gyroscope is pointed to the north at the start of a plane or boat trip. A motor inside the gyrocompass keeps it spinning and it will stay pointed to the north despite the speed that the plane or boat is traveling, rough seas, or air currents (**turbulence**). Using the cord and following your teacher's directions, start your gyroscope. When it is spinning rapidly, place it on its frame and move the gyroscope and frame next to either compass. Align the axis of the spinning gyroscope with the compass needle so that its top is pointing to the north. Now, pick up the frame and spinning gyroscope, and walk around the room with it as you did with the compass in Step 8. Note in which direction the top of the gyroscope is pointing. Record your answer in the Data Collection and Analysis section.

Putting the Compass to Use

10. You will now learn how to use a field compass. For this learning exercise, we will assume that the magnetic north pole and the geographic north pole coincide and that there is no magnetic deviation.

Look at your field compass, which is also called an **orienteering compass**. Compare it to the diagram below. Notice in particular that its **housing** moves independently of the card. The housing has degree readings on it. Observe that north, west, south, and east are each 90 degrees apart and that the full circle is 360 degrees. Locate the **direction-of-travel arrow** indicated on the base of the compass (Figure 3).

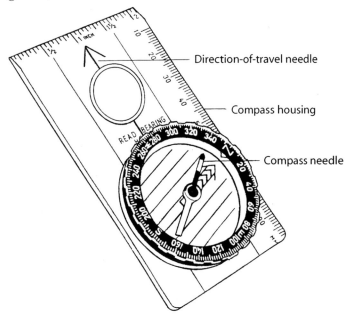

Figure 3: Field (orienteering) compass

(continued)

Finding Your Way Using a Magnet *(continued)*

An empty baseball diamond is a good place to learn. When you get on the field, go to home plate and do the following:

(a) Face first base. Hold the compass so that it is level with the ground.

(b) Turn the base of the compass so that the direction-of-travel arrow is pointing directly at first base.

(c) Turn the compass housing until the red (arrowhead) end of the compass needle is pointing at the "N." Now, read the direction to first base. It is indicated by the direction-of-travel needle. Record that direction in the Data Collection and Analysis section. **CAUTION: Be sure the north-seeking end of the compass needle is pointing to the "N." If the south-seeking end of the compass needle is pointing to the "N," you will be reading the opposite direction!**

(d) Go to first base. Find the direction from first to second base. Record that direction in the Data Collection and Analysis section.

(e) What is the direction from second base to third? Record that direction in the Data Collection and Analysis section.

(f) What is the direction from third base to home plate? Record that direction in the Data Collection and Analysis section.

(g) Suppose you wanted to walk west from home plate. How would you do it? Record your answer in the Data Collection and Analysis section.

DATA COLLECTION AND ANALYSIS

1. Why must you remove all iron and steel materials from the work table on which the compasses are placed? _____

_____

2. Describe what happened to the magnetized needle on the cork when the unit was placed on the water in the saucer. _____

_____

3. How do the directions indicated on the field compass and on your homemade compass compare when they are placed on the tabletop? _____

_____

4. The end of the needle that pointed to the north magnetic pole of the earth was the _____

5. The reason(s) why a field compass would be unsatisfactory for navigating a jet plane or rapidly moving boat is (are) _____

_____

(continued)

Finding Your Way Using a Magnet *(continued)*

6. *Optional*—As you walked around the room with the spinning gyroscope, what were your observations? _____

_____

Position	Direction (NW, SE, etc.)	Degrees
Home to first base		
First to second base		
Second base to third		
Third to home		

7. Describe how to use the compass to travel in a westerly direction from home plate. _____

_____

❔ CONCLUDING QUESTIONS

1. Why is knowing the magnetic deviation in your area important for hikers?_____

_____

2. Describe how you can use a field compass to travel in a southeast direction from your home.

_____

_____

〰 Follow-up Activities 〰

1. Use a navigational map to find the magnetic deviation in your area. How can you use that information to change the results you got on the ball field? After showing your work to your teacher, change your answers in the table above. Share your information with your classmates.

2. Research how GPS (Global Positioning System) works. Prepare a report for your teacher.

3. Find out how electronic compasses work. Report your findings to your class.

Exploring Magnetic Fields

✔ INSTRUCTIONAL OBJECTIVES

Students will be able to

- draw conclusions based on observations.
- demonstrate the presence of a magnetic field around a magnet.
- state and demonstrate the law of magnetic poles.

🌐 NATIONAL SCIENCE STANDARDS ADDRESSED

Students demonstrate a conceptual understanding of

- transformations of energy.
- big ideas and unifying concepts such as cause and effect.
- motion and forces such as net forces and magnetism.

Students demonstrate scientific inquiry and problem-solving skills by

- evaluating the outcomes of the investigation.
- working in teams to collect and share information and ideas.
- identifying and controlling experimental variables.

Students demonstrate effective scientific communication by

- arguing from evidence such as data collected through their own experimentation.
- explaining scientific concepts to other students.

Students demonstrate competence with the tools and technologies of science by

- using tools and technology to collect data.

✂ MATERIALS

- Two bar magnets
- String
- Test-tube clamp
- Support stand
- Field compass
- Piece of white chalk
- Two packages of small metal paper clips
- Iron filings sealed inside a transparent plastic case *or* glass plate and iron fillings in a salt shaker
- Piece of cardboard, approximately 20 cm by 25 cm

HELPFUL HINTS AND DISCUSSION

Time frame: One class period
Structure: Individually or in cooperative learning groups of two students
Location: In class

In this activity, students will affirm the law of magnetic poles, demonstrate a magnetic field surrounding a bar magnet, and investigate the inclination of a compass needle. It is best to use iron filings that are enclosed in a transparent plastic case because loose iron filings on a glass plate are difficult to collect after the activity has been completed. Demonstrate to your students how to look for the inclination of the compass needle in the last part of the activity.

ADAPTATIONS FOR HIGH AND LOW ACHIEVERS

High Achievers: Stress the need for accuracy while collecting the data in this activity. Encourage these students to carry out the Follow-up Activities. They should also work with and assist the low achievers in the cooperative learning groups, especially in setting up the suspended magnet and observing the angle of inclination of the compass needle.

Low Achievers: Organize these students in cooperative learning groups with the high achievers. Discuss the need for cooperation among the members of each group. Provide a glossary and reference material for boldfaced terms.

SCORING RUBRIC

Full credit should be given to those students whose predictions, rationales, observations, and answers are complete and accurate. Students should answer the Concluding Questions in complete sentences. Give extra credit to those students who complete the Follow-up Activities.

 **INTERNET TIE-INS**   http://www.magneticfield.com
http://www.tchnicoil.com/magnetism.html
http://www.exploratorium.edu/snacks/magnetic_lines.html

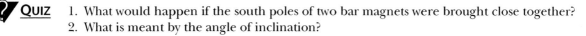 **QUIZ**   1. What would happen if the south poles of two bar magnets were brought close together?
2. What is meant by the angle of inclination?

Name _____ Date _____

Exploring Magnetic Fields

⚡ BEFORE YOU BEGIN ⚡

Magnets play an important role in our lives. They are found in many electrical devices, including audiospeakers, motors, and doorbells. Magnets attract objects made of iron or steel, such as paper clips. The most familiar magnets are ones that hold notes to the refrigerator door.

All magnets have two ends, called poles: a north pole and a south pole. The forces that magnets exert on one another are similar to the forces between positive and negative electric charges. Like poles (N-N, S-S) repel and unlike poles (N-S) attract each other. This statement is termed **the law of magnetic poles**. In this activity, you will confirm that this law is correct.

A magnet also has a magnetic field around it. This invisible field runs from pole to pole. In this activity you will make this field visible. You will also determine the effect a magnetic field has on a sensitive detector of magnetic fields, namely, a compass needle.

MATERIALS

- Two bar magnets
- String
- Test-tube clamp
- Support stand
- Field compass

- Piece of white chalk
- Two packages of small metal paper clips
- Iron filings sealed inside a transparent plastic case *or* glass plate and iron filings in a salt shaker
- Piece of cardboard, approximately 20 cm by 25 cm

PROCEDURE

1. Your first task is to be sure that the law of magnets is correct. Tie one end of the string around the middle of one of the magnets so that when you lift it by the string it is perfectly balanced.

2. Suspend the magnet from the test-tube clamp attached to a support stand and allow the magnet to swing about freely (Figure 1). Remove any iron or steel objects from the desk top. Make sure to use a desk top made of wood or other nonmetallic material.

Figure 1

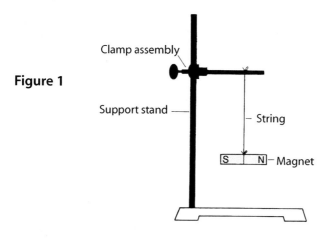

(continued)

Exploring Magnetic Fields *(continued)*

3. Observe how the magnet aligns itself. Place the standard compass on the table as far as possible from the bar magnet. How does its alignment compare with that of the bar magnet? Record your answer in the Data Collection and Analysis section.

4. With the chalk, mark an "N" at the end of the magnet pointing north and an "S" at the end of the magnet pointing south.

5. Repeat Steps 1–4 for the other bar magnet.

6. Bring the ends of the magnets marked "N" toward each other. Record your observations in the table in the Data Collection and Analysis section.

7. Repeat Step 6 for the two ends marked "S."

8. Bring the end marked "N" of one magnet toward the end marked "S" of the other magnet and again record your observations in the table in the Data Collection and Analysis section.

9. Place one of the bar magnets in the center of the piece of cardboard. Sprinkle the paper clips over the magnet. Where do most of the paper clips end up? Record your observations in the Data Collection and Analysis section.

10. Remove the paper clips from the magnet and return the magnet to its place on the cardboard. Pick up the plastic case with the iron filings. Hold the case in the horizontal position and gently shake it in order to disperse the filings evenly throughout. Now, place the plastic case carefully over the magnet. NOTE: If you do not have the plastic case containing the iron filings, place the glass plate over the magnet and shake the iron filings over the area covering the magnet. What happened to the iron filings? Record your observations in the Data Collection and Analysis section. In the space provided in the Data Collection and Analysis section, draw a picture of the pattern the iron filings formed. This pattern illustrates the **magnetic lines of force** that comprise the magnetic field.

11. Leave the plastic case or glass plate over the magnet. Slowly bring the compass over the north pole of the magnet, and rest it on the plastic or glass. Bend down and observe the compass needle's vertical position. Record your observation in the Data Collection and Analysis section.

12. Repeat Step 11, but place the compass over the south pole, then over the center of the magnet. The angle made is called the **angle of inclination**. *Keep the angle of inclination in mind because it will be referred to in a future activity.*

DATA COLLECTION AND ANALYSIS

1. How did the alignment of the suspended bar magnet compare to the compass needle? _____

_____

Magnet Position	Observation
"N" to "N"	
"S" to "S"	
"N" to "S"	

(continued)

Exploring Magnetic Fields *(continued)*

2. Where did most of the paper clips end up after you sprinkled them on the magnet? _____
   _____

3. Where were the fewest paper clips found after you sprinkled them on the magnet? _____
   _____

4. Describe what happened to the iron filings. _____
   _____

Diagram of Magnetic Lines of Force

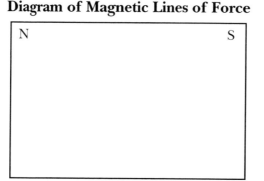

5. Describe the angle of inclination of the compass needle at the north pole. _____
   _____

6. Describe the angle of inclination of the compass needle at the south pole._____
   _____

7. Describe the angle of inclination of the compass needle at the center of the magnet. _____
   _____

CONCLUDING QUESTIONS

1. Where are magnetic lines of force the strongest?_____

2. If we think of the earth as having a giant bar magnet buried in its core, where would a compass
   needle be inclined the most?_____ The least? _____

⋙ Follow-up Activities ⋙

1. Determine the magnetic field around a horseshoe magnet, a ring magnet, a magnetic
   marble.

2. Research and prepare a report for your teacher dealing with the possible healing powers
   of magnetic fields.

3. Investigate magnetic fields, sunspots, and the aurora borealis on the Internet and in the
   library. Share your findings with your classmates.

The Simplest of Circuits: Series Circuits

✔ INSTRUCTIONAL OBJECTIVES

Students will be able to

- explain how the various components of a series circuit are connected.
- draw conclusions based on observations.
- construct simple series electrical circuits.

🌐 NATIONAL SCIENCE STANDARDS ADDRESSED

Students demonstrate a conceptual understanding of

- electrical motions or transformations of energy.
- big ideas and unifying concepts such as cause and effect.

Students demonstrate scientific inquiry and problem-solving skills by

- identifying the problem and evaluating the outcomes of the investigation.
- working in teams to collect and share information and ideas.
- recognizing, analyzing, and critiquing alternative explanations.
- identifying and controlling experimental variables.

Students demonstrate effective scientific communication by

- arguing from evidence such as data collected through their own experimentation.
- explaining scientific concepts to other students.

Students demonstrate competence with the tools and technologies of science by

- using tools and technology to collect data.

✂ MATERIALS

- Three flashlight bulbs (3-volt ratings)
- Two dry-cell batteries (6-volt standard lantern batteries)
- Three flashlight bulb holders
- Five 25-cm lengths of insulated copper wire with both ends stripped of insulation for approximately 2 cm and alligator clips attached to each end
- Switch

HELPFUL HINTS AND DISCUSSION

Time frame: One class period
Structure: Cooperative learning groups of three students
Location: In class

In this activity, students will learn how series circuits can be built and how they work. Point out to the students that electricity flows through a complete circuit, which is a continuous, unbroken loop. Discuss the components of a circuit: a source of electricity (dry cell battery), electrical wires, resistors (flashlight bulbs), and a switch. Mention the fact that electrons moving along a wire create an electric current. Encourage the students to predict before experimenting. Organize the class into cooperative learning teams. Stress the need to check the connections in each circuit assembled. Test the batteries and bulbs beforehand to be sure that they are in good working condition.

ADAPTATIONS FOR HIGH AND LOW ACHIEVERS

High Achievers: Stress the need for as much accuracy as possible in collecting the data in this activity. Encourage these students to carry out the Follow-up Activities. They should also work with and assist the low achievers in the cooperative learning groups.

Low Achievers: Organize these students in cooperative learning groups with the high achievers. Discuss the need for cooperation among the members of each group. Provide a glossary and reference material for boldfaced terms.

SCORING RUBRIC

Full credit should be given to those students whose predictions, rationales, observations, and answers are complete and accurate. Students should answer the Concluding Questions in complete sentences. Give extra credit to those students who complete the Follow-up Activities.

 **INTERNET TIE-INS**
http://www.iit.edu/~smile/ph8917.html
http://www.rpsec.usca.sc.edu/RPSEC/Student/tsp/elec.html

QUIZ
1. What components are necessary to construct a series circuit?
2. What will happen if the connections are not properly attached?
3. What occurs when the switch in the series circuit is closed?
4. Name three electrical appliances that operate using a series circuit.

Name _____ Date _____

The Simplest of Circuits: Series Circuits

⚡ BEFORE YOU BEGIN ⚡

In an **electric circuit**, electricity flows in an unbroken path from a power source to an **electrical device** and then back to the power source. For example, when you switch on a flashlight, you complete a loop that enables the electricity to flow from the dry-cell batteries, which are the power source. The electricity then travels to the flashlight bulb, which is the electrical device, and then back to the battery (Figure 1). Once all are connected, the light bulb glows. In order for electricity to flow, all the parts of the simple circuit must be connected to each other. If there is a break anywhere in the path of the electricity, the electricity cannot flow. The **switch** on a flashlight is used to complete or break the path of the electricity.

Several things are required to make a simple circuit: the power source (a dry-cell battery), electrical wires (which allow the electricity to flow along easily), a switch (which can complete or interrupt the circuit), and **resistors**.

Resistors—in this activity, light bulbs—can be connected in several ways. In this activity, they will be connected in a **series circuit**. In a series circuit, all the parts are connected to one another in a single path, like three people holding hands in a circle. In this activity, you will construct several series circuits.

✂ MATERIALS

- Three flashlight bulbs (3-volt ratings)
- Two dry-cell batteries (6-volt standard lantern batteries)
- Three flashlight bulb holders

- Five 25-cm lengths of insulated copper wire with both ends stripped of insulation for approximately 2 cm and alligator clips attached to each end
- Switch

⬦ PROCEDURE

PART I—SERIES CIRCUITS WITH A SINGLE BULB

Use the diagrams below in Figure 1 as a guide to set up your circuits. Before constructing the circuits, predict, in each case, if your team thinks the flashlight bulb will glow or not. Record your predictions in Table 1 in the Data Collection and Analysis section.

Figure 1: Series (simple) circuits—one bulb

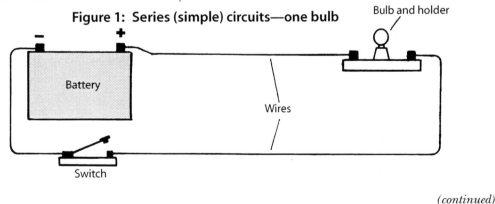

(continued)

The Simplest of Circuits: Series Circuits *(continued)*

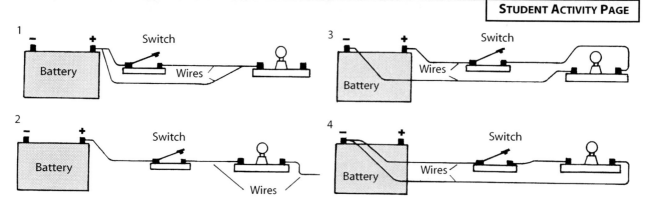

1. Obtain from your teacher a dry-cell battery, three copper wires with alligator clips on the ends, a switch, a flashlight bulb, and a bulb holder. Locate the positive and negative **terminals** on the battery.

2. Have your group predict if the bulb will light or not. Record the prediction in Table 1 in the Data Collection and Analysis section.

3. Set up the first circuit illustrated in "Series Circuit—One Bulb" by attaching the alligator clips on the wires to the terminals on the battery and the light bulb holder. Close the switch. Did the bulb light? Was your prediction correct? Record your results in Table 1 in the Data Collection and Analysis section.

4. Repeat Steps 2 and 3 for the second circuit.

5. Repeat Steps 2 and 3 for the third and fourth circuits.

PART II—SERIES CIRCUITS WITH TWO BULBS

Use the diagrams below in Figure 2 as a guide to set up your circuits. Before constructing the circuits, predict, in each case, if your team thinks the flashlight bulbs will glow or not. Record your predictions in Table 2 in the Data Collection and Analysis section.

Figure 2: Series circuits—two bulbs

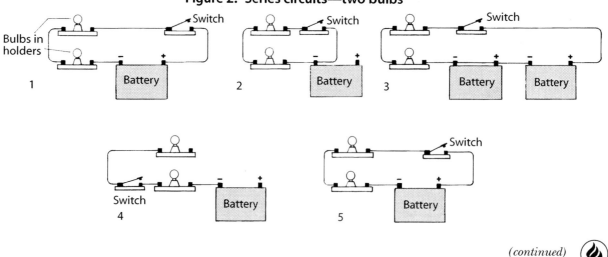

(continued)

Name ____ Date ________________________

The Simplest of Circuits: Series Circuits *(continued)*

STUDENT ACTIVITY PAGE

1. Obtain from your teacher a dry-cell battery, five copper wires with alligator clips on the ends, a switch, two flashlight bulbs, and two bulb holders. Locate the positive and negative **terminals** on the battery.
2. Remember to have your group predict if one bulb or both bulbs will light or not. Record the prediction in Table 2 in the Data Collection and Analysis section.
3. Set up the first circuit illustrated in the diagrams "Series Circuit—Two Bulbs" by attaching the alligator clips on the wires to the terminals on the battery and the light bulb holder. Close the switch. Did both bulbs light? Did only one light? Was your prediction correct? If both bulbs glowed, remove one bulb. What happened to the other bulb? Record your results in the Data Collection and Analysis section for Series Circuits—Two Bulbs.
4. Repeat Steps 2 and 3 for the second circuit.
5. Repeat Steps 2 and 3 for the third, fourth, and fifth circuits.

DATA COLLECTION AND ANALYSIS

TABLE 1: SERIES (SIMPLE) CIRCUIT—ONE BULB

Circuit	Prediction	Observation after Closing the Switch	Was the Prediction Correct?
1			
2			
3			
4			

1. In which circuits did the flashlight bulb glow? ____________________________________

____

2. What do these circuits have in common?____

____

____

3. Hypothesize why the other circuits failed to work.___________________________________

____

____

(continued)

The Simplest of Circuits: Series Circuits *(continued)*

TABLE 2: SERIES CIRCUIT—TWO BULBS

Circuit	Prediction	Observation—Bulbs Lit?	Was the Prediction Correct?
1			
2			
3			
4			
5			

1. In which circuits did both light bulbs glow? _____

2. In which circuits did only one bulb glow? _____

3. In which circuits did neither light bulb glow? _____
   Why do you think this happened? _____

4. In circuits in which both bulbs glowed, what happened to the other bulb when one bulb was removed from the circuit? _____

5. How did the intensity of the lights in circuit #3 compare to the intensity of the lights in circuit #1? _____

❓ CONCLUDING QUESTIONS

1. Why does the light intensity change as you add more bulbs to the series circuit? _____
   _____

2. If Christmas lights on a string are connected to each other in series, what will happen to the rest if one bulb burns out? _____
   _____

3. Based on your observations, what is the disadvantage of a series circuit? _____
   _____

4. Under what circumstances would this disadvantage be an advantage? _____
   _____

⚡ Follow-up Activities ⚡

1. Design your own series circuit and predict how brightly each bulb will glow. You can use two batteries and as many bulbs as you wish. Write down your prediction and have your team test your hypothesis.

2. Take a flashlight apart and determine how the circuit is completed.

If It Isn't a Series Circuit, It Must Be a Parallel Circuit!

✔️ INSTRUCTIONAL OBJECTIVES

Students will be able to

- explain how the various components of a parallel circuit are connected.
- draw conclusions based on observations.
- construct parallel electrical circuits.

🌐 NATIONAL SCIENCE STANDARDS ADDRESSED

Students demonstrate a conceptual understanding of

- electrical motions or transformations of energy.
- big ideas and unifying concepts such as cause and effect.

Students demonstrate scientific inquiry and problem-solving skills by

- identifying the problem and evaluating the outcomes of the investigation.
- working in teams to collect and share information and ideas.
- recognizing, analyzing, and critiquing alternative explanations.
- identifying and controlling experimental variables.

Students demonstrate effective scientific communication by

- arguing from evidence such as data collected through their own experimentation.
- explaining scientific concepts to other students.

Students demonstrate competence with the tools and technologies of science by

- using tools and technology to collect data.

✂️ MATERIALS

- Three flashlight bulbs (3-volt ratings)
- Two dry-cell batteries (6-volt standard lantern batteries)
- Three flashlight bulb holders
- Ten 25-cm lengths of insulated copper wire with both ends stripped of insulation for approximately 2 cm and alligator clips attached to each end
- Switch

HELPFUL HINTS AND DISCUSSION

Time frame: One class period
Structure: Cooperative learning groups of three students
Location: In class

In this activity, students will learn how to construct parallel circuits. It is important that your students carry out the series circuit activity before attempting this activity. Point out to the students the difference between a series circuit and a parallel circuit. Review the components of a circuit: a power source (dry-cell battery), electrical wires, resistors (flashlight bulbs), and a switch. Mention the fact that electrons moving along a wire create an electric current. Encourage the students to predict before testing each circuit. Organize the class into cooperative learning teams. Stress the need to check the connections in each circuit assembled. Test the batteries and bulbs beforehand to be sure that they are in good working condition.

ADAPTATIONS FOR HIGH AND LOW ACHIEVERS

High Achievers: Stress the need for as much accuracy as possible in collecting the data in this activity. Encourage these students to carry out the Follow-up Activities. They should also work with and assist the low achievers in the cooperative learning groups.

Low Achievers: Organize these students in cooperative learning groups with the high achievers. Discuss the need for cooperation among the members of each group.

SCORING RUBRIC

Full credit should be given to those students whose predictions, rationales, observations, and answers are complete and accurate. Students should answer the Concluding Questions in complete sentences. Give extra credit to those students who complete the Follow-up Activities.

 **INTERNET TIE-INS**   http://www.iit.edu/~smile/ph9113.html
http://gamstcwebgisd.k12.mi.us/msta/journal/misconceptions-spring98.html
http://www.matter.org.uk/schools/Content/Resistors/Default.htm

QUIZ   1. Why are Christmas tree lights wired in parallel?
2. In what ways do series and parallel circuits differ?

If It Isn't a Series Circuit, It Must Be a Parallel Circuit!

⚡ BEFORE YOU BEGIN ⚡

In a previous activity, you learned how to connect light bulbs, battery, and switch in a series circuit. The electric current had only one path to travel through the circuit. By doing this activity, you will learn how to construct a **parallel circuit**. In a parallel circuit, each bulb is connected directly to the battery, as shown in Figure 1 in the Procedure section diagram. You will also learn how parallel and series circuits differ. Finally, you will determine if one bulb in the circuit gets more current than the others. Just bear in mind that the more current a bulb gets, the brighter the light.

✂ MATERIALS

- Three flashlight bulbs (3-volt ratings)
- Two dry-cell batteries (6-volt standard lantern batteries)
- Three flashlight bulb holders

- Ten 25-cm lengths of insulated copper wire with both ends stripped of insulation for approximately 2 cm and alligator clips attached to each end
- Switch

🔧 PROCEDURE

Use the diagrams in Figure 2 below as a guide to set up your circuits. Before constructing the circuits, predict, in each case, if your team thinks that one, two, or three flashlight bulbs will glow or not. Record your predictions in Table 1 in the Data Collection and Analysis section.

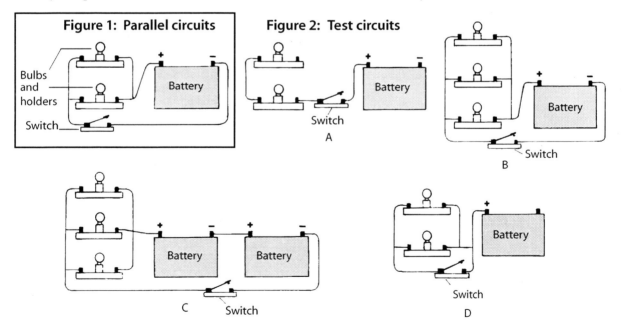

Figure 1: Parallel circuits

Figure 2: Test circuits

(continued)

If It Isn't a Series Circuit, It Must Be a Parallel Circuit! *(continued)*

1. Obtain two dry-cell batteries, ten copper wires with alligator clips on the ends, a switch, three flashlight bulbs, and three bulb holders from your teacher. Locate the positive and negative **terminals** on the battery.

2. Remember to have your group predict which bulbs, if any, will light. Record the prediction in Table 1 in the Data Collection and Analysis section.

3. Set up the first circuit illustrated in Parallel Circuits by attaching the alligator clips on the wires to the terminals on the battery and the light bulb holder. Close the switch. Did the bulbs light? Was your prediction correct? Record your results in Table 1 in the Data Collection and Analysis section.

4. If both bulbs glowed, remove one bulb from its holder. What was the effect of this action upon the second bulb? Record your answer in the Data Collection and Analysis section.

5. Repeat Steps 2, 3, and 4 for the second, third, and fourth circuits.

DATA COLLECTION AND ANALYSIS

TABLE 1

Circuit	Prediction	Observation	Was the Prediction Correct?
A			
B			
C			
D			

1. In which circuits did all of the flashlight bulbs glow?_____

_____

2. What do these circuits have in common?_____

_____

3. Hypothesize why the other circuits failed to work. _____

_____

4. In which circuit did the bulbs glow the brightest? _____

5. How can you explain any difference in intensity? _____

_____

6. What happened to the intensity of the light in the bulbs in Circuit C?_____

_____

7. What happened when you removed a bulb in those circuits in which all the bulbs glowed?

_____

_____

(continued)

If It Isn't a Series Circuit, It Must Be a Parallel Circuit! *(continued)*

? CONCLUDING QUESTIONS

1. Why do some bulbs glow less brightly in some of the parallel circuits? _____

   _____

   _____

2. Why are parallel circuit wiring schemes used in homes and businesses? _____

   _____

   _____

3. Based on your observations of series and parallel circuits, what are the advantages of parallel circuits? _____

   _____

   _____

⚡ Follow-up Activities ⚡

1. Design your own parallel circuit and predict how brightly each bulb will glow. You can use two batteries and as many bulbs as you wish. Write down your prediction and have your team test your hypothesis.

2. Research the types of circuits used with:
   circuit breakers
   alarm and security systems
   outdoor lights

Fuses, Circuit Breakers, and Heat

✓ INSTRUCTIONAL OBJECTIVES

Students will be able to

- explain the action of fuses and circuit breakers as electrical safety devices.
- draw conclusions based on observations.
- construct simple models of fuses and thermostatic elements in circuit breakers.

🌐 NATIONAL SCIENCE STANDARDS ADDRESSED

Students demonstrate a conceptual understanding of

- electrical motions or transformations of energy.
- big ideas and unifying concepts such as cause and effect.

Students demonstrate scientific inquiry and problem-solving skills by

- identifying the problem and evaluating the outcomes of the investigation.
- working in teams to collect and share information and ideas.
- recognizing, analyzing, and critiquing alternative explanations.
- identifying and controlling experimental variables.

Students demonstrate effective scientific communication by

- arguing from evidence such as data collected through their own experimentation.
- explaining scientific concepts to other students.

Students demonstrate competence with the tools and technologies of science by

- using tools and technology to collect data.

✂ MATERIALS

- Dry-cell battery (6 volts—fully charged)
- Flashlight bulb in bulb holder
- Five 25-cm lengths of insulated copper wire with both ends stripped of insulation for approximately 2 cm
- Alligator clips
- Switch
- Steel wool (preferably without soap)
- Heat resistant pad
- Spool of 20-gauge nickel-chromium wire (chromel)
- Scissors
- Strip of brass and invar (unequal expansion bar), 10 cm in length
- Metal forceps
- Small piece of lens paper
- Small container of water

HELPFUL HINTS AND DISCUSSION

Time frame: One class period
Structure: Cooperative learning groups
Location: In class

In this activity, students will learn how fuses and circuit breakers protect against fires and overheating in wires and other electrical equipment. Caution the students not to touch the nickel-chromium wire with their hands when it is in the circuit. Tell them to allow sufficient time for it to cool before removing it from the circuit. Have a small container of water on hand to put out the burning paper.

Test the unequal expansion bar beforehand. Tell the students to place one side (the side that will bend *upward* when heated) facing them. You might color code the sides with small stickers so they don't make a mistake.

ADAPTATIONS FOR HIGH AND LOW ACHIEVERS

High Achievers: Encourage these students to carry out the Follow-up Activities. They also should work with the low achievers in the cooperative learning groups while setting up the circuits and answering the questions.

Low Achievers: Organize these students in cooperative learning groups with the high achievers. Discuss the need for cooperation among the members of each group. Provide a glossary and reference material for boldfaced terms in this activity.

SCORING RUBRIC

Full credit should be given to those students whose predictions, rationales, observations, and answers are complete and accurate. Students should answer the Concluding Questions in complete sentences. Give extra credit to those students who complete the Follow-up Activities.

INTERNET TIE-INS   http://www.programma.se/matcb/html
http://ourworld.compuserve.com/homepages/g_knott/elect16.htm
http://atschool.eduweb.co.uk/trinity/projects/fusetest/fuse.html

QUIZ   1. In a fuse box, the following was written:
Do not use silver paper or nails!
Always replace the fuse with the correct value!
You have been warned!
Explain what could happen if these cautions were ignored.

2. How are fuses and circuit breakers similar?

3. How are fuses and circuit breakers different?

Fuses, Circuit Breakers, and Heat

⚡ BEFORE YOU BEGIN ⚡

If there is a fault in a wire or a piece of equipment in an electric circuit, then an unusually large amount of electricity may flow through the circuit. This could cause the wire or equipment to over-heat and possibly start a fire. Fuses and circuit breakers prevent this from happening. A fuse is a piece of wire in either a glass or a ceramic container. The piece of wire is the main part. It will carry a specific amount of electricity ($\frac{1}{2}$ **amp**, 10 amps, etc.). If the current in the circuit goes above that amount, the fuse wire melts, creating a gap in the circuit. No current can flow until the fuse is replaced. Fuses usually carry a bit more current than the equipment that they are protecting requires. For example, if the equipment requires 1.75 amps, its fuse would carry 2 amps. You should never replace a melted fuse until you have found and repaired the cause of the overload. And you should never replace a fuse with one that will carry more current. Always replace the melted fuse with one of the recommended amount after you have corrected the cause of the melt-ing of the old fuse.

Circuit breakers do the same thing as fuses. But a circuit breaker is a type of electrical switch that opens automatically when the current goes above a certain amount. A circuit breaker can be reset, while a fuse must be replaced after it has opened. Circuit breakers are opened by a device such as an electromagnet that can detect an unusually high current, or by a temperature-sensitive device such as a **thermostat**.

The first thing you will do in this activity is to see how electricity can cause a fire. You will also make a fuse and see how it protects a circuit from overheating. Your final task will be to make a model of a thermostat that could be inside of a circuit breaker.

✄ MATERIALS

- Dry-cell battery (6 volts—fully charged)
- Flashlight bulb in bulb holder
- Five 25-cm lengths of insulated copper wire with both ends stripped of insulation for approximately 2 cm
- Alligator clips
- Switch
- Steel wool (preferably without soap)
- Heat resistant pad

- Spool of 20-gauge nickel-chromium wire (chromel)
- Scissors
- Strip of brass and invar (unequal expansion bar), 10 cm in length
- Metal forceps
- Small piece of lens paper
- Small container of water

◈ PROCEDURE

Use the diagram in Figure 1 that follows as a guide to set up your circuit. Record all answers to the questions in the Data Collection and Analysis section.

(continued)

Fuses, Circuit Breakers, and Heat *(continued)*

Figure 1

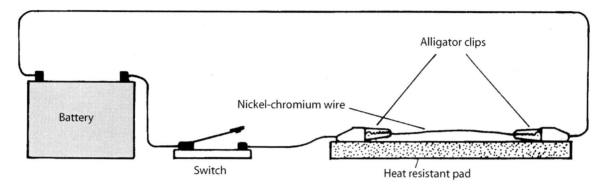

ELECTRICITY CAN CREATE HEAT

1. Using the scissors, cut a piece of nickel-chromium wire that will fit across the top of the heat resistant pad.

2. Set up the circuit as shown above (Figure 1). Securely clip the nickel-chromium wire strip between both alligator clips. Be sure that the strip of metal makes good contact with the alligator clips.

3. Tighten all connections. Close the switch and observe the nickel-chromium strip.

4. When you see a significant change in the appearance of the wire strip, pick up a small piece of lens paper with the forceps. Touch the paper to the wire strip. (If the paper catches fire, douse the burning paper in the container of water.) What happened to the paper? How can you explain your observation?

FUSE

Use the diagram that follows in Figure 2 as a guide to set up your circuit. Record all answers to the questions in the Data Collection and Analysis section.

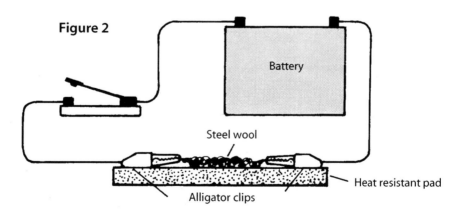

Figure 2

(continued)

Fuses, Circuit Breakers, and Heat *(continued)*

1. Start by removing one thin strand of steel wool from the steel wool pad. Rest it on top of the heat resistant pad.

2. Set up the circuit as shown on the previous page (Figure 2). Carefully attach the alligator clips to the ends of the steel wool strand. Don't break the strand of steel wool!

3. Close the switch and observe the strand of steel wool. Record your observation in the Data Collection and Analysis section.

4. Open the switch. Replace the steel wool strand, if necessary. Insert a bulb in a bulb holder in this series circuit between the switch and the steel wool strand.

5. Close the switch and observe both the strand and the bulb. Record both observations.

6. Open the switch. Replace the steel wool strand, if necessary. Insert a bulb in a bulb holder. This time, create a parallel circuit as shown in the diagram below in Figure 3.

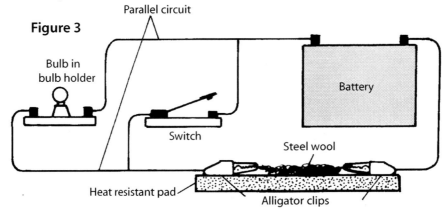

Figure 3

Parallel circuit

Bulb in bulb holder

Battery

Switch

Steel wool

Heat resistant pad

Alligator clips

7. Does the bulb light with the switch open? Why? Record your answer and the answers to the questions that follow in the Data Collection and Analysis section.

8. Does the fuse (strand of steel wool) melt when the switch is open? _____ Explain. _____

_____

9. Close the switch. What happens to the bulb and to the fuse (strand of steel wool)?_____

_____

THERMOSTAT

1. Remove the bulb from the parallel circuit that you just made. Insert the bulb in the series circuit as in Step 4 above (Figure 2).

2. Remove the steel wool strand from the alligator clips. Attach one alligator clip to one end of the piece of unequal expansion bar. Ask your teacher which side of the bar should be facing upward. Rest the other end of the unequal expansion bar *on top of the other alligator clip.* Be sure the correct side is facing upward.

3. Close the switch. Wait a few minutes. What happened to the unequal expansion bar? Wait about five minutes more. Now, what happened to the expansion bar? Record your answer in the appropriate space.

(continued)

Fuses, Circuit Breakers, and Heat *(continued)*

4. Repeat Step 3. How do your observations compare to the first time you carried out Step 3?

   _____

DATA COLLECTION AND ANALYSIS

1. What happened to the nickel-chromium wire when you closed the switch? _____

   _____

2. What happened to the lens paper? _____

   How can you explain this observation? _____

   _____

3. What happened to the steel wool strand when you closed the switch? _____

   _____

4. What happened to the "fuse" (steel wool strand) when you closed the switch in the series circuit?

   _____

   What happened to the bulb? _____

5. What happened to the "fuse" (steel wool strand) when you left the switch open in the parallel circuit? _____

   What happened to the bulb? _____

   Explain this observation. _____

   _____

6. What happened to the "fuse" (steel wool strand) when you closed the switch in the parallel circuit? _____

   What happened to the bulb? _____

   Explain this observation. _____

   _____

7. What happened to the unequal expansion bar during the first few minutes? _____

   _____

8. What happened to the unequal expansion bar during the next few minutes? _____

   _____

9. How do your results the second time you carried out Step 3 compare with your original observations?

   _____

   _____

(continued)

Fuses, Circuit Breakers, and Heat *(continued)*

❓ CONCLUDING QUESTIONS

1. Why is the action of the unequal expansion bar similar to the action of a circuit breaker? _____
   _____
   _____

2. Why are fuses and circuit breakers important? _____
   _____
   _____

3. What advantage do circuit breakers have over fuses? _____
   _____
   _____

4. It is easy to see if a glass fuse has melted because you can see the wire inside. Ceramic fuses are covered with solid opaque material. How can you tell if this type of fuse has melted inside?
   _____
   _____

〰 Follow-up Activities 〰

1. Research **antisurge** fuses and report on how they work.

2. Research how circuit breakers are tested to be sure they are functioning properly.

3. Suppose you used a piece of thin copper wire instead of a strand of steel wool in these experiments. What do you think would happen when you carried out Steps 3 through 9 in the "Fuse" section of the Procedure? Test your hypotheses if your teacher approves.

Conductors vs. Insulators

✓ INSTRUCTIONAL OBJECTIVES

Students will be able to

- explain the difference between a conductor and an insulator.
- draw conclusions based on observations.

🌐 NATIONAL SCIENCE STANDARDS ADDRESSED

Students demonstrate a conceptual understanding of

- electrical motions or transformations of energy.
- big ideas and unifying concepts such as cause and effect.

Students demonstrate scientific inquiry and problem-solving skills by

- identifying the problem and evaluating the outcomes of the investigation.
- working individually and in teams to collect and share information and ideas.
- recognizing, analyzing, and critiquing alternative explanations.

Students demonstrate effective scientific communication by

- arguing from evidence such as data collected through their own experimentation.
- explaining scientific concepts to other students.

Students demonstrate competence with the tools and technologies of science by

- using tools and technology to collect data.

✂ MATERIALS

- 6-volt lantern battery
- Three insulated copper wires, approximately 15 cm in length
- 6-volt bulb in bulb holder
- Ball of string
- Plastic ruler
- Piece of paper
- Paper clip
- Piece of wood
- Piece of aluminum foil
- Rubber eraser
- Piece of wool cloth
- Piece of silk cloth
- Piece of cotton cloth
- Stainless steel fork
- Glass microscope slide
- Piece of gold jewelry
- Silver coin or silver jewelry
- Small, unused plastic drinking cup
- Distilled water
- Salt in a salt shaker
- Glass stirring rod

HELPFUL HINTS AND DISCUSSION

Time frame: One class period
Structure: Cooperative learning groups
Location: In class

In this activity, students test various materials to determine whether the materials are conductors or insulators. Point out to the students that conductors have many free electrons that can easily conduct electricity. Insulators do not have many free electrons and therefore do not conduct electricity. Demonstrate how to hold the insulated portions of the wires when testing the items.

ADAPTATIONS FOR HIGH AND LOW ACHIEVERS

High Achievers: Have these students carry out the Follow-up Activities. They also should work with and assist the low achievers in the cooperative learning groups.

Low Achievers: Organize these students in cooperative learning groups with the high achievers. Provide a glossary and reference material for boldfaced terms in this activity.

SCORING RUBRIC

Full credit should be given to those students whose predictions, rationales, observations, and answers are complete and accurate. Students should answer the Concluding Questions in complete sentences. Give extra credit to those students who complete the Follow-up Activities.

INTERNET TIE-INS   http://www.cs.stewards.edu/wright/sci20/conduct.html
http://www.projectgcse.co.uk/physics/conductors.htm
http://www.mos.org/sln/toe/simpleelectroscope.html

QUIZ   1. Why are some materials conductors while others are insulators?
2. Name three good conductors of electricity and three insulators.

Name _____   Date _____

Conductors vs. Insulators

⚡ BEFORE YOU BEGIN ⚡

Everything you see and touch is made up of tiny particles called **atoms**. Even **elements**, such as copper, oxygen, and carbon, are constructed of atoms. An atom is the smallest part of an element that has all the properties of that element. For example, the smallest part of copper is one atom of copper.

All atoms are made up of even smaller particles called **protons**, **electrons**, and **neutrons**. The **nucleus** of an atom contains two types of particles: protons, which are positively charged, and neutrons, which have no charge. Spinning around the nucleus like planets orbiting the sun are the negatively charged electrons, as seen in Figure 1 below.

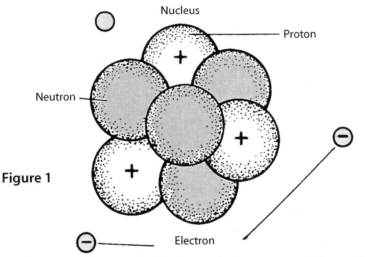

Figure 1

The attraction between the negatively charged electrons and the positively charged protons helps hold the atom together. As with magnets, some attractions are stronger than others. In elements where the attraction between the electrons and protons is weak, the electrons are able to leave the atoms and move throughout the material. This movement of electrons from atom to atom is electricity. Elements in which the attraction is weak and the electrons flow easily are called **conductors**. They can easily conduct electricity. Copper and aluminum are examples of conductors.

On the other hand, some materials have electrons that are tightly bound around the nucleus. Thus, the flow of electrons is, for practical purposes, zero. These are **insulators**, such as rubber and glass. These materials are poor conductors of electricity. All materials can be classified as either conductors or insulators, based on their ability to conduct electricity. In this activity, you will classify various materials as either conductors or insulators.

(continued)

Name _____ Date _____

Conductors vs. Insulators *(continued)*

STUDENT ACTIVITY PAGE

MATERIALS

- 6-volt lantern battery
- Three insulated copper wires, approximately 15 cm in length
- 6-volt bulb in bulb holder
- Ball of string
- Plastic ruler
- Piece of paper
- Paper clip
- Piece of wood
- Piece of aluminum foil
- Rubber eraser

- Piece of wool cloth
- Piece of silk cloth
- Piece of cotton cloth
- Stainless steel fork
- Glass microscope slide
- Piece of gold jewelry
- Silver coin or silver jewelry
- Small, unused plastic drinking cup
- Distilled water
- Salt in a salt shaker
- Glass stirring rod

PROCEDURE

1. Connect one pole of the bulb holder to the negative pole of the dry cell using one of the wires, as shown in Figure 2 below.

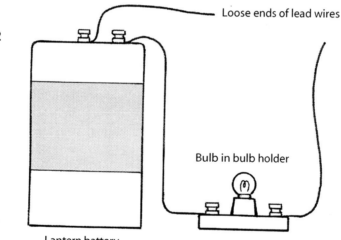

Figure 2

Loose ends of lead wires

Bulb in bulb holder

Lantern battery

2. Connect another wire to the other pole of the bulb holder.
3. Connect the third wire to the positive end of the lantern battery.
4. The two loose ends of wire will serve as the test terminals. First, test the battery's strength by bringing the two loose ends together. Keep your fingers on the insulated part of the wire when testing the battery and all other materials.
5. Before testing, predict which items you believe are insulators and which are conductors by putting an "I" or a "C" in the prediction column in the Data Collection and Analysis section.

(continued)

© 2000 J. Weston Walch, Publisher   58   *Walch Hands-on Science Series: Electricity and Magnetism*

Conductors vs. Insulators *(continued)*

6. Test the solid materials for conductivity by placing each test item between the two loose wires. Then, *while holding the insulated portion of the wires*, bring the bare ends in contact with the item. Be sure that the two loose ends do not touch each other. If the bulb lights, the item is a conductor. If the bulb fails to light, the item is an insulator. Record your results by placing an "X" in the proper box in the Data Collection and Analysis section.

7. Fill the plastic glass half full with distilled (deionized) water. Predict if the distilled water is a conductor or insulator, then test by inserting the two wire ends into the water. Record your results by placing an "X" in the proper box in the Data Collection and Analysis section.

8. Add a small amount of salt to the water in the glass from the salt shaker. Stir the solution with the stirring rod. Predict if the salt water is a conductor or insulator, then test by inserting the two wire ends into the water. Record your results by placing an "X" in the proper box in the Data Collection and Analysis section.

DATA COLLECTION AND ANALYSIS

Test Item	Prediction	Conductor	Insulator
String			
Wood			
Aluminum foil			
Rubber eraser			
Wool			
Rayon			
Cotton			
Paper clip			
Stainless steel fork			
Glass microscope slide			
Piece of gold			
Piece of silver			
Piece of paper			
Plastic ruler			
Distilled (deionized) water			
Salt water			

(continued)

Name _____   Date _____

STUDENT ACTIVITY PAGE

1. What percentage of your predictions was correct? _____

2. How can you explain the conductivity of distilled water and of salt water?_____

   _____

   _____

❓ CONCLUDING QUESTIONS

1. What are the characteristics of insulators and conductors? _____

   _____

   _____

2. Gold is an excellent conductor of electricity, yet is rarely used. Why? _____

   _____

   _____

⚡ Follow-up Activities ⚡

1. Research how superconductors work and explain why they would be ideal as electricity supply cables. Report your findings to the class.

2. Investigate how conductive paints are made and how they are used. Prepare a report on this topic for your teacher.

Producing Electricity from Electrochemical Cells

✔️ INSTRUCTIONAL OBJECTIVES

Students will be able to

- explain how an electrochemical cell produces electricity.
- draw conclusions based on observations.
- construct a variety of electrochemical cells.

🌐 NATIONAL SCIENCE STANDARDS ADDRESSED

Students demonstrate a conceptual understanding of

- electrical motions or transformations of energy.
- big ideas and unifying concepts such as cause and effect.

Students demonstrate scientific inquiry and problem-solving skills by

- identifying the problem and evaluating the outcomes of the investigation.
- working individually to collect and share information and ideas.
- recognizing, analyzing, and critiquing alternative explanations.

Students demonstrate effective scientific communication by

- arguing from evidence such as data collected through their own experimentation.
- explaining scientific concepts to other students.

Students demonstrate competence with the tools and technologies of science by

- using tools and technology to collect data.

✂️ MATERIALS

- Strip of copper
- Strip of zinc
- Two insulated wires approximately 30 cm in length, bared at each end
- Two alligator clips
- Screwdriver
- Two 500 cc beakers
- Copper sulfate solution
- Zinc sulfate solution
- Potassium nitrate solution
- U-tube
- Grapefruit
- Orange
- Lemon
- Banana
- 1.5-volt bulb in socket
- Glass wool to plug the U-tube

HELPFUL HINTS AND DISCUSSION

Time frame: One class period
Structure: Individual
Location: In class

In this activity, students learn how electricity is created in an electrochemical cell. Demonstrate how to set up the salt bridge using glass wool (which is a nonconductor) to seal the ends. It would be wise to attach the alligator clips to one end of each wire before the class begins. Instruct the students on the proper way to hook up the wires to the voltmeter. Check the connections made by the students.

ADAPTATIONS FOR HIGH AND LOW ACHIEVERS

High Achievers: Have these students carry out the Follow-up Activities. They also should help the low achievers construct the electrochemical cells.

Low Achievers: Provide these students with a glossary and reference material for boldfaced terms in this activity. Also provide them with detailed instructions regarding their tasks.

SCORING RUBRIC

Full credit should be given to those students whose observations and answers are complete and accurate. Students should answer the Concluding Question in a complete sentence. Give extra credit to those students who complete the Follow-up Activities.

INTERNET TIE-INS   http://www.virtualsoftware.com/prodpage.cfm?ProdID=667
http://www.naio.kcc.hawaii.edu/chemistry/everyday_electro.html

QUIZ   1. What are the components of an electrochemical cell?
2. Where do the electrons that flow through the wire in the lemon electrochemical cell come from?

Producing Electricity from Electrochemical Cells

⚡ BEFORE YOU BEGIN ⚡

An **electrochemical cell** is a device that produces electricity by changing chemical energy into electrical energy. An electrochemical cell consists of two different metal strips called **electrodes** connected by wires to a voltmeter or some electrical device. The metal strips are placed in a conducting solution called an **electrolyte**. Look at the electrochemical cell at right in Figure 1.

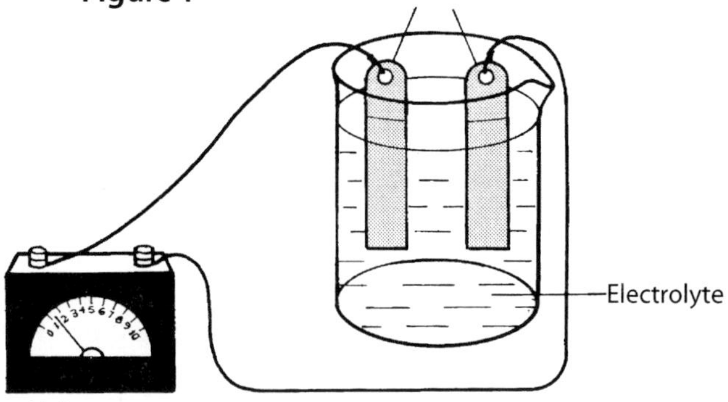

One metal in an electrochemical cell will lose electrons and the other metal will gain electrons. A chemical reaction occurs. This forces electrons to flow along a wire from the electrode that loses electrons to the electrode that gains electrons. For example, look at the electrochemical cell at right (Figure 2).

In this cell, electrons leave the zinc strip. They travel along the wire to the copper strip. Along the way, they pass through the voltmeter in the middle. The zinc strip is in a **zinc sulfate solution**. The copper strip is in a **copper sulfate solution**. When the zinc and copper strips are connected by a wire, a chemical change begins. Zinc atoms in the strip of zinc enter the zinc sulfate solution. As each zinc atom enters the solution, it leaves two electrons behind. This means that the zinc atoms are no longer electrically neutral. Each atom now has a positive charge because of losing two negative electrons. The remaining part of the zinc atom is called an **ion**. The two electrons that were left behind move through the wire to the copper strip, creating an electric current in the wire. This can be seen on the voltmeter that is attached to the wire. Now positively charged copper ions that are in the copper sulfate solution react with those electrons to form copper atoms. The salt bridge is a U-tube containing a salt solution that allows the ions to travel through it. Without the salt bridge, electricity would not flow. Electrochemical cells act like pumps to force electrons to flow around a circuit. This force is known as **voltage**. **Batteries** are composed of two or more electrochemical cells hooked up to each other so that they combine their electric currents. The more electrochemical cells in a battery, the greater its power.

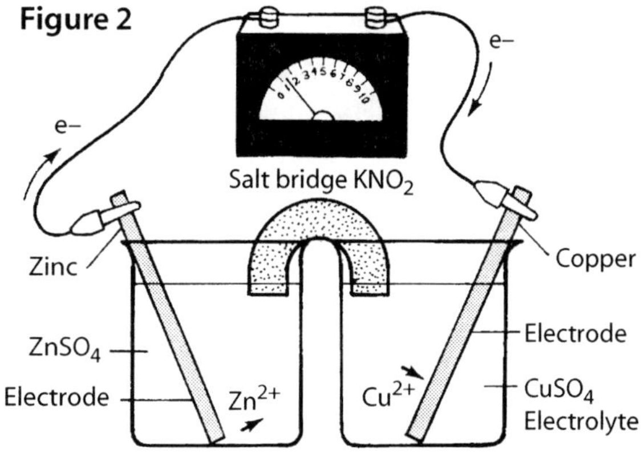

In this activity, you will first make an electrochemical cell as shown above and measure its voltage. Then, you will make electrochemical cells from lemons and other fruits and determine their voltages.

Figure 1

Electrodes

Electrolyte

Voltmeter

Voltmeter

Figure 2

e−

e−

Salt bridge KNO$_2$

Zinc

Copper

ZnSO$_4$

Electrode

Zn^{2+} Cu^{2+}

Electrode

CuSO$_4$
Electrolyte

(continued)

Name _____ Date _____

Producing Electricity from Electrochemical Cells *(continued)*

MATERIALS

- Strip of copper
- Strip of zinc
- Two insulated wires approximately 30 cm in length, bared at each end
- Two alligator clips
- Screwdriver
- Two 500 cc beakers
- Copper sulfate solution
- Zinc sulfate solution

- Potassium nitrate solution
- U-tube
- Grapefruit
- Orange
- Lemon
- Banana
- 1.5-volt bulb in socket
- Glass wool to plug the ends of the U-tube

PROCEDURE

1. Use the diagram in Figure 2 on the previous page to set up your electrochemical cell. Your teacher will show you how to fill and use the salt bridge. Using the screwdriver, attach the alligator clips to each wire as shown in Figure 2. Record the voltage you read from the voltmeter in the table in the Data Collection and Analysis section.

2. Replace the voltmeter with the bulb and connect the wires to it. Does it light? Write your answer in the table in the Data Collection and Analysis section.

3. Set up an electrochemical cell using a lemon as shown below (Figure 3). Stick the copper and zinc strips in the lemon as shown. Record the voltage you read from the voltmeter in the table in the Data Collection and Analysis section.

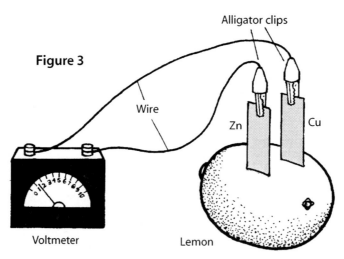

Figure 3

Alligator clips

Wire

Zn   Cu

Voltmeter   Lemon

4. Replace the voltmeter with the bulb. Again, determine if the bulb lights and record your answer in the table in the Data Collection and Analysis section.

5. Repeat Steps 3 and 4, substituting an orange, grapefruit, and banana for the lemon.

(continued)

Producing Electricity from Electrochemical Cells *(continued)*

DATA COLLECTION AND ANALYSIS

Electrochemical Cell	Voltage	Bulb (Lit/Did Not Light)
Chemicals		
Lemon		
Orange		
Grapefruit		
Banana		

1. What were the electrolytes in each of the fruits tested?_____
   _____

2. Why is a lemon an example of an electrochemical cell? _____
   _____

3. Which fruit produced the highest voltage? _____

4. Which fruit produced the lowest voltage? _____

5. What reason can you offer for a particular fruit yielding little or no voltage? _____
   _____

CONCLUDING QUESTION

How does a lemon produce electricity when zinc and copper strips are inserted in it? _____
_____
_____

⚡ Follow-up Activities ⚡

1. Determine how long a 1.5-volt bulb will keep on glowing when connected to a lemon electrochemical cell.

2. Substitute a tomato for a lemon. How do your results compare?

3. Research what would happen if the first electrochemical cell you made did not have a salt bridge.

4. Construct a battery by using two lemons. How does the voltage compare to that produced by a single lemon? Explain why there is a difference. Report your results to your classmates.

Creating Electricity from Ice Water

✔️ INSTRUCTIONAL OBJECTIVES

Students will be able to

- explain how a thermocouple works.
- draw conclusions based on observations.
- construct a thermocouple.

🌐 NATIONAL SCIENCE STANDARDS ADDRESSED

Students demonstrate a conceptual understanding of

- electrical motions or transformations of energy.
- motion and forces such as electricity.
- big ideas and unifying concepts such as cause and effect.

Students demonstrate scientific inquiry and problem-solving skills by

- identifying the problem and evaluating the outcomes of the investigation.
- working in teams to collect and share information and ideas.
- using physical science concepts to explain observations.
- identifying and controlling experimental variables.

Students demonstrate effective scientific communication by

- arguing from evidence.
- explaining scientific concepts to other students.

Students demonstrate competence with the tools and technologies of science by

- using a thermometer.
- using a voltmeter.

✂️ MATERIALS

- 500-ml beaker
- Voltmeter
- 🖐️ Gas burner
- 50-cm length of iron or aluminum wire
- Two copper wires, each 25 cm in length
- 🖐️ Matches
- Ice cubes
- Water
- Support stand
- Test-tube clamp with plastic coated ends
- Celsius thermometer

🖐️ = Safety icon

HELPFUL HINTS AND DISCUSSION

Time frame: One class period
Structure: Cooperative learning groups
Location: In class

In this activity, your students will make and use a thermocouple to create electricity. The thermocouple is made by joining two dissimilar wires at each end. When the ends are in different temperature environments, an electric current will be produced. Thermocouples normally produce voltages in the range between 7 and 20 millivolts. Be sure to advise your students which posts of the voltmeter should be used for connecting the copper wires. Supervision of the lighting of the gas burner and the placement of the "hot end" of the thermocouple should be supervised by you or another knowledgeable adult. Make the students aware that a hot wire and a cool one look exactly alike; care must be taken when handling the "hot end" of the thermocouple.

ADAPTATIONS FOR HIGH AND LOW ACHIEVERS

High Achievers: Discuss the importance of controlling the temperature in the design of this activity. Encourage these students to carry out the Follow-up Activities. They should work with the low achievers in the cooperative learning groups.

Low Achievers: Discuss the safety hazards in this activity. Organize these students in cooperative learning groups with the high achievers. Provide a glossary for the terms in boldface.

SCORING RUBRIC

Full credit should be given to those students who record observations and correctly answer all questions using full sentences. Give extra credit for completing the Follow-up Activities.

**INTERNET TIE-INS**   http://www.xco.com/How_it_works.htm
http://www.iqinstruments.com/temperature/thermo.html
http://www.saunderstech.com/basics.htm

QUIZ   1. What is a thermocouple?
2. How can a thermocouple be used to show the temperature of an automobile engine?

Creating Electricity from Ice Water

⚡ BEFORE YOU BEGIN ⚡

In this activity you will produce electricity using ice water and heat. The electrical device you will build is called a **thermocouple** and the electricity produced in this manner is called **thermo-electricity.** The diagram in Step 1 of the procedure is a thermocouple. You will also determine the relationship between the number of **millivolts** (1/1000's of a volt) and the temperature difference between the "hot" and "cold" ends of the thermocouple.

✂ MATERIALS

- 500-ml beaker
- Voltmeter
- ✋ Gas burner
- 50-cm length of iron or aluminum wire
- Two copper wires, each 25 cm in length
- ✋ Matches

- Ice cubes
- Water
- Support stand
- Test-tube clamp with plastic coated ends
- Celsius thermometer

✋ = Safety icon

📦 PROCEDURE

1. Use the diagram in Figure 1 below as a guide to set up your thermocouple.

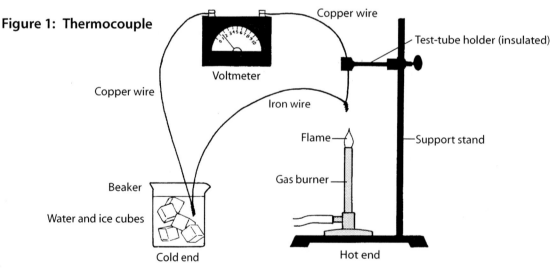

Figure 1: Thermocouple

Copper wire

Test-tube holder (insulated)

Copper wire

Iron wire

Flame

Support stand

Gas burner

Voltmeter

Beaker

Water and ice cubes

Cold end

Hot end

2. Attach one end of each piece of copper wire to the voltmeter terminals that were designated by your teacher.

3. Attach the iron wire to the free end of each of the copper wires by twisting both wires around each other. Complete six turns as shown in the diagram above.

4. Set up the gas burner and ring stand as shown. Do not ignite the burner at this time.

(continued)

Creating Electricity from Ice Water *(continued)*

5. **THIS STEP MUST BE DONE UNDER THE DIRECT SUPERVISION OF YOUR TEACHER.** Ignite the burner and under your teacher's direction, setting the flame for maximum heat. On the upright arm of the support stand, make a mark in pencil or pen at the height of the top of the inner cone of the gas flame. Turn off the gas.

6. Position the end of the thermocouple wires closest to the ring stand just above the mark you made in the previous step. Hold the end in position by placing it in the test tube holder clamp about 20 cm from the end that will be in the flame. Tighten the clamp. Be sure the clamp end is not directly over the flame but away from it, otherwise the insulated ends will burn.

7. Fill the beaker with water at room temperature.

8. Insert the free ("cold") end of the thermocouple into the water. Now insert the thermometer into the beaker of water.

9. **THIS STEP MUST BE DONE UNDER THE DIRECT SUPERVISION OF YOUR TEACHER.** Relight the burner. If necessary, change the position of the "hot" end of the thermocouple so that it is just above the inner cone of the flame. Do this by adjusting the test tube clamp.

10. Take your initial reading of the thermometer and of the voltmeter. Record them in the data table in the Data Collection and Analysis section. We will assume for all readings that the temperature of the hot end of the thermocouple is 1500 degrees Celsius.

11. Add ice cubes. When the temperature drops four degrees, take another reading of the voltmeter.

12. Repeat Step 11 until you have a minimum of four readings. Five or six would be better.

13. Complete the data table.

DATA COLLECTION AND ANALYSIS

Temperature— Hot End	Temperature— Cold End	Temperature Difference	Number of Millivolts
1500° C			
1500° C			
1500° C			
1500° C			
1500° C			
1500° C			

1. At which temperature for the "cold" end was the number of millivolts the greatest?

_____

(continued)

Creating Electricity from Ice Water *(continued)*

2. From your data table, what is the relationship between the temperature difference of the "hot" and "cold" ends and the number of millivolts produced? _____
_____
_____

3. Why is thermoelectricity a good name for this process? _____
_____
_____

❓ CONCLUDING QUESTIONS

1. How can a thermocouple be used to measure the heat of the engine in a car? _____
_____
_____

2. Many modern gas stoves contain thermocouples. Why? _____
_____
_____

⚡ Follow-up Activities ⚡

1. Investigate other uses of thermocouples, including as safety devices, by using the Internet.

2. Research the use of thermocouples made of chromel and alumel as well as those made of platinum and rhodium. Share your findings with the class.

3. Repeat this activity, but attach the meter to the iron or aluminum wire rather than the copper wires. How do your results compare with your original experiment?

Producing Electricity:
The Thirty-three Cent Battery and Photovoltaics

✔ INSTRUCTIONAL OBJECTIVES

Students will be able to

- compare an electrochemical cell battery to a dry cell.
- draw conclusions based on observations.
- construct a battery consisting of three electrochemical cells.
- discuss the advantages of photovoltaic electricity.
- demonstrate practical uses for photovoltaic cells and panels.

🌐 NATIONAL SCIENCE STANDARDS ADDRESSED

Students demonstrate a conceptual understanding of

- electrical motions or transformations of energy.
- big ideas and unifying concepts such as cause and effect.
- interactions of energy and matter.
- the impact of technology.

Students demonstrate scientific inquiry and problem-solving skills by

- identifying the problem and evaluating the outcomes of the investigation.
- working individually to collect and share information and ideas.
- recognizing, analyzing, and critiquing alternative explanations.

Students demonstrate effective scientific communication by

- arguing from evidence such as data collected through their own experimentation.
- explaining scientific concepts to other students.

Students demonstrate competence with the tools and techniques of science by

- using tools and technology to collect data.

✂ MATERIALS

- Two insulated wires approximately 30 cm in length, bared at each end
- Strong detergent

- Stiff cleaning brush
- Three shiny pennies
- Three shiny dimes
- One quarter (25-cent piece)
- Small alligator clips
- Coil of thin copper wire
- Scissors
- Thin sponge
- 400-ml beaker containing a concentrated salt solution (NaCl)
- *Optional*—floodlight (for use if the day is overcast)
- One or two photovoltaic cells (each should generate up to 300 mA and 0.5 v)
- 75-watt bulb in a bulb holder
- Milliampere meter
- Small battery-operated radio
- Galvanometer
- LED or small neon bulb
- Forceps
- Paper toweling
- Screwdriver
- Tape (duct/masking)

HELPFUL HINTS AND DISCUSSION

Time frame: One class period
Structure: Cooperative learning groups
Location: In class

Explain to the students that there are many ways of producing electricity and that they are going to work with two of them. First, the students will learn how electricity is created in an electrochemical battery consisting of three dimes and three pennies. They will determine the effect of increasing and decreasing the number of cells upon the output of current and voltage. In the second half of this activity they will use photovoltaic (solar) cells and panels to operate electrical devices. Demonstrate how to connect and read the galvanometer, voltmeter, and milliampere meter.

ADAPTATIONS FOR HIGH AND LOW ACHIEVERS

High Achievers: Have these students carry out the Follow-up Activities. They should help the low achievers construct the electrochemical cells, particularly in regard to the taping and wiring.

Low Achievers: Provide a glossary and reference material for boldfaced terms in this activity. These students should be included in groups with high achievers.

SCORING RUBRIC

Full credit should be given to those students whose observations and answers are complete and accurate. Students should answer the Concluding Questions in complete sentences. Give extra credit to those students who complete the Follow-up Activities.

 INTERNET TIE-INS   http://www.virtualsoftware.com/prodpage.cfm?ProdID=667
http://www.naio.kcc.hawaii.edu/chemistry/everyday_electro.html
http://eren.doe.gov/pv/howworks.html
http://www.freeyellow.com/members6/vandewater/sol_proj.html

QUIZ   1. An ordinary dry cell contains carbon and zinc and moist ammonium chloride. What parts of the coin battery are the equivalents of the three materials found in the dry cell?

2. You are going to explore part of the desert in the western United States. State three reasons that it would be wise to include solar panels as part of your equipment.

Producing Electricity: The Thirty-three Cent Battery and Photovoltaics

⚡ BEFORE YOU BEGIN ⚡

There are many ways to produce electricity. One is with a **battery** that produces electricity by changing chemical energy into electrical energy. All batteries are composed of two or more **electrochemical cells** hooked up to each other. As a result, their electric currents are combined. The battery that you will construct in this exercise has three cells. All electrochemical cells have two **electrodes**, each a different kind of metal. Your electrochemical cells will have silver electrodes made from dimes. The other electrodes will be made using copper pennies. Finally, you need an **electrolyte**, which is necessary to transfer the electrons. In your coin battery, the electrolyte is salt water. Without the electrolyte, no electricity would be produced.

Another way of producing electricity is with a **photovoltaic cell**. This changes light energy into electricity. After completing the second part of this activity, devoted to photoelectric cells, you can tell your family that you know a way to have the electric meter in your home run *backwards!* That would happen if the solar cells on your roof were connected to the electrical system powering your home.

Only the light energy *absorbed* by **PV** (photovoltaic) cells can be changed into electricity. Light that is reflected off the PV cell cannot be converted to electricity. Likewise, any light that passes through the PV cell will not be changed into electricity. PV cells are very simple. They have no moving parts and require no fuel. A PV cell has a **transparent** cover, usually glass. Inside are two kinds of **semiconductors**, with different electrical properties. These semiconductors are similar to those found in **computers**. The two semiconductors touch each other. When the semiconductors are placed in the light, an electric current is produced. Scientists link a number of PV cells together to form a **panel**. A panel of PV cells is similar to a battery formed by linking a number of electrochemical cells. In this activity, you will form and use a panel of PV cells.

✂ MATERIALS

- Two insulated wires approximately 30 cm in length, bared at each end
- Strong detergent
- Stiff cleaning brush
- Three shiny pennies
- Three shiny dimes
- One quarter (25-cent piece)
- Small alligator clips
- Coil of thin copper wire
- Scissors
- Thin sponge

- 400-ml beaker containing a concentrated salt solution (NaCl)
- *Optional*—floodlight (for use if the day is overcast)
- One or two photovoltaic cells (each should generate up to 300 mA and 0.5 v)
- 75-watt bulb in a bulb holder
- Milliampere meter
- Small battery-operated radio
- Galvanometer
- **LED** or small neon bulb

(continued)

Producing Electricity: The Thirty-three Cent Battery and Photovoltaics *(continued)*

- Forceps
- Paper toweling
- Screwdriver
- Tape (duct/masking)

PROCEDURE

The Thirty-three Cent Battery

1. Using the brush and detergent solution, scrub the three dimes and pennies until they shine and are free of dirt and oxidized material.

2. Using the quarter as a template, cut out five round pieces of the sponge.

Figure 1

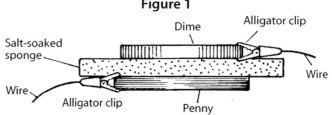

3. Place the pieces of sponge in the salt-water solution. Using the forceps or your hands, be sure each piece is saturated with salt solution.

4. Prepare three cells, each one consisting of a dime and a penny separated from each other by a piece of sponge that has been saturated with salt water. Attach the small alligator clips to the wires with the screwdriver. Then, attach the wires to the coins as shown in Figure 1 above.

5. Assemble the three cells as shown in Figure 2 below. In order to assemble the three cells into a battery, you will have to wrap them with tape. Dry the top and bottom coins with the paper toweling so that the tape will stick to them. Wrap the three cells together using the duct or masking tape. While wrapping the cells together, don't wrap them too tight. A snug wrap is fine.

Figure 2

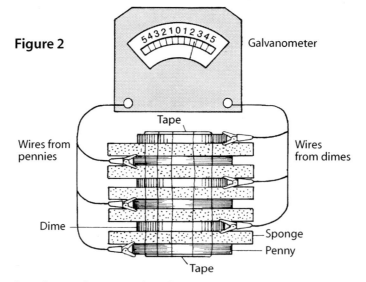

Just don't squeeze the pieces of sponges hard enough to lose the electrolyte solution (salt water) in the sponges. Compare your assembly with Figure 2 above.

(continued)

74

Producing Electricity: The Thirty-three Cent Battery and Photovoltaics *(continued)*

6. Connect the galvanometer to the battery as shown in Figure 2. Record the galvanometer reading in the Data Collection and Analysis section.

7. Replace the galvanometer with an LED or small neon bulb. Record your observations in the Data Collection and Analysis section.

8. Repeat Steps 6 and 7 using only one cell rather than three. Record your observations in the Data Collection and Analysis section.

Photovoltaics

Record all answers in the Data Collection and Analysis section.

1. Be sure you are working in a sunny part of the room or have floodlights available.

2. Connect one solar cell to the voltmeter. Place the solar cell so that it is directly in the path of the light, either sunlight or artificial light. What is the voltage reading on the meter?

3. Remove the batteries from the radio. Attach the solar cell leads to the radio's battery terminals. Does it play? If not, try reversing the leads from the solar cell. If it still doesn't play, try using two solar cells in series attached to the radio.

4. In similar fashion, attach one, then two, etc., solar cells to the 75-watt light bulb. How many were required to make it light?

DATA COLLECTION AND ANALYSIS

1. What was the galvanometer reading for the three-cell battery? _____
_____

2. What did you observe when you replaced the galvanometer with an LED or bulb? _____
_____

3. What was the galvanometer reading for one cell? _____
_____

4. What was your observation of the light source using one cell?_____
_____

5. What was the voltage produced by the solar cell?_____

6. How many solar cells were required to play the battery-operated radio? _____

7. How many were required to light the 75-watt bulb?_____

(continued)

Producing Electricity: The Thirty-three Cent Battery and Photovoltaics *(continued)*

❓ CONCLUDING QUESTIONS

1. How can you increase the voltage and current of a coin battery? _____
   _____

2. Was the galvanometer reading using one cell one third that of the three-cell reading? _____
   What is a possible reason for your answer? _____
   _____

3. Why are PV cells essential to our space program? _____
   _____

〰 Follow-up Activities 〰

1. Prepare and test similar coin batteries using other coins, such as silver dollars or foreign coins. As a result of your experimentation, generalize about the conditions that will increase the amount of voltage and current produced by a coin battery.

2. Research the many advantages of PV cells and panels. Report your results to your classmates.

3. Research how DC current (current flowing in one direction) produced by solar cells can be converted to AC current (current that reverses itself at regular intervals). Prepare a news article for your school newspaper on this topic.

Piezoelectricity: Electricity with a Twist and Semiconductors

✔ INSTRUCTIONAL OBJECTIVES

Students will be able to

- describe the generation of electrical energy by mechanical stress on piezoelectric crystals.
- state the uses of piezoelectricity.
- draw conclusions based on observations.
- explain how a semiconductor differs from a conductor and an insulator.

🌐 NATIONAL SCIENCE STANDARDS ADDRESSED

Students demonstrate a conceptual understanding of

- electrical motions or transformations of energy.
- big ideas and unifying concepts such as cause and effect.

Students demonstrate scientific inquiry and problem-solving skills by

- identifying the problem and evaluating the outcomes of the investigation.
- working individually and in teams to collect and share information and ideas.
- recognizing, analyzing, and critiquing alternative explanations.

Students demonstrate effective scientific communication by

- arguing from evidence such as data collected during their own experimentation.
- explaining scientific concepts to other students.

Students demonstrate competence with the tools and techniques of science by

- using tools and technology to collect data.

✂ MATERIALS

- Percussion hammer
- LED/glow lamp
- Two pieces of wire, 20 cm in length, with alligator clips on each end
- #2 pencil (prepared by you, with part of the wood cut away to expose the graphite core)
- 6-volt lantern battery
- 3.8-volt bulb in bulb holder
- Voltmeter
- Piezoelectric cell
- Metric rule

HELPFUL HINTS AND DISCUSSION

Time frame: One class period
Structure: Cooperative learning groups
Location: In class

In this activity, your students will create electricity using a piezoelectric crystal in a cell. They will also set up and use a semiconductor device. It would be best if you cut the wood away along the length of the pencil, as shown in the diagram below. You might demonstrate a piezoelectric gas

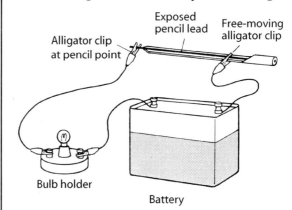

or barbecue lighter as a motivational device for this activity. Semiconductors are included in this activity because of their importance to the functioning of computers and many other electrical and electronic devices. The photovoltaic cells that are explored in Activity 13 in this book could not function if there were no semiconductors. Point out to your students that both topics in this activity are independent of each other and both are important.

77

ADAPTATIONS FOR HIGH AND LOW ACHIEVERS	SCORING RUBRIC
High Achievers: These students should carry out the Follow-up Activities. Have these students assist the low achievers, particularly when measuring distances in the semiconductor part of this activity. **Low Achievers:** Provide a glossary and reference material for boldfaced terms in this activity. Encourage these students to work with the high achievers.	Full credit should be given to those students whose observations and answers are complete and accurate. Students should answer the Concluding Questions in complete sentences. Give extra credit to those students who complete the Follow-up Activities.

 **INTERNET TIE-INS**

http://www.piezo.com/histry.html
http://met121.bham.ac.uk/6ulkpzt.htm
http://ourworld.compuserve.com/homepages/boyce_smith/diodes.htm

 QUIZ

1. What is piezoelectricity?
2. What materials are piezoelectrical in nature?
3. Name two uses for semiconductors.

Piezoelectricity: Electricity with a Twist and Semiconductors

⚡ BEFORE YOU BEGIN ⚡

More than a hundred years ago, scientists discovered that crystals of quartz and cane sugar would produce electricity if they were twisted or squeezed in certain directions. Producing electricity in this way was named **piezoelectricity**. "Piezo" means "pressure." Some ceramic materials will also produce electricity when they are compressed.

You have seen piezoelectricity at work in many devices without realizing it. Your family may use a barbecue lighter that uses no fuel or batteries and never wears out. By moving the switch on the handle of the lighter, sparks are produced by a piezoelectric crystal inside. Radio transmitters use quartz crystals in order to work.

In this activity, you will research what happens to a piezoelectric cell when it is squeezed, bent, or hit with a hammer. You will also study a type of **semiconductor**. Semiconductors are materials that resist the flow of **electrons**, which is what electricity is. Semiconductors permit *less* electricity to pass through them than **conductors** do. Conductors do not resist the passage of electricity. Semiconductors allow *more* electricity to pass through them than do **insulators**. Insulators are so resistant to the passage of electricity that no electrons flow. **Silicon** is a common material used to make semiconductors. Semiconductors are found in may electronic devices, including **LEDs**, or **light-emitting diodes**.

✂ MATERIALS

- Percussion hammer
- LED/glow lamp
- Two pieces of wire, 20 cm in length, with alligator clips on each end
- #2 pencil (prepared by your teacher, with part of the wood cut away to expose the graphite core)

- 6-volt lantern battery
- 3.8-volt bulb in bulb holder
- Voltmeter
- Piezoelectric cell
- Metric rule

📐 PROCEDURE

1. Connect the LED or the neon bulb directly to the piezoelectric cell, if possible. If not, use the wires with the alligator clips. Record your observations for the following steps in Table 1 in the Data Collection and Analysis section. Look for a flash of light in the light source.

2. Hold the cell in both hands, being careful not to touch the terminals or bare wires. Try to bend the cell gently.

3. Repeat Step 2, but this time bend the piezoelectric cell as hard as you can.

4. Squeeze, but do not bend, the piezoelectric cell gently.

5. Squeeze, but do not bend, the piezoelectric cell as hard as you can.

(continued)

Piezoelectricity: Electricity with a Twist and Semiconductors *(continued)*

6. Use the hammer and tap the cell lightly.

7. Using the hammer, tap the piezoelectric cell harder. Try not to hit the cell so hard that it would break.

8. Replace the light with the voltmeter. Repeat Steps 2 through 7 and record any voltages you obtain in the table in the Data Collection and Analysis section.

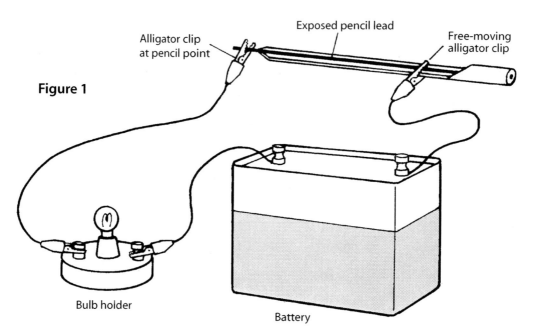

Figure 1

Alligator clip at pencil point
Exposed pencil lead
Free-moving alligator clip
Bulb holder
Battery

9. Now you will work with graphite, which is a semiconductor of electricity. The "lead" in a lead pencil is really graphite. Look at Figure 1 above and set up the equipment as shown.

10. Handle the wires by holding onto the insulation. First bring both alligator clips together so that they touch each other. Record whether the bulb lights or not in Table 2 in the Data Collection and Analysis section. If the bulb does light, state whether it is glowing brightly or dimly.

11. Keep the fixed alligator clip attached to the pencil point. Slowly move the free-moving alligator clip away from the fixed clip until the bulb does not glow. Using the metric rule, measure the distance from that point to the fixed point. Record your measurement in millimeters in Table 2.

12. Divide the measurement recorded in Step 11 by 2. Place the free-moving alligator clip at that point. Describe the intensity of the light bulb in Table 2 in the Data Collection and Analysis section.

13. Replace the light bulb with the voltmeter and repeat Steps 10 through 12. Report the voltages in Table 2 in the Data Collection and Analysis section.

(continued)

Piezoelectricity: Electricity with a Twist and Semiconductors *(continued)*

DATA COLLECTION AND ANALYSIS

TABLE 1

Action	Observation of Light Source	Voltmeter Reading
Gentle bending		
Strong bending		
Gentle squeeze		
Strong squeeze		
Light tap		
Hard tap		

TABLE 2

Settings	Light (Description)	Distance from Fixed Clip	Voltmeter Reading
Both alligators together		Zero	
Point at which bulb no longer lights	- - - - - - - - - -		Zero
Halfway point			

1. Which action on the piezoelectric cell produced the largest amount of electricity? _____

   _____

2. Which action produced the least amount of electricity? _____

   _____

3. How is the semiconductor you just set up similar to a dimmer switch for an electric light?

   _____

(continued)

Piezoelectricity: Electricity with a Twist and Semiconductors *(continued)*

❓ CONCLUDING QUESTIONS

1. From your data and your answers to the questions above, what generalization can you make about the relationship between stress and the amount of electricity produced? _____
_____
_____

2. How could a piezoelectric device be used as a sensing device for mechanical stress? _____
_____
_____

3. Which has more resistance to electricity, copper wire or graphite?
_____

Explain your answer._____
_____
_____

⚡ Follow-up Activities ⚡

1. Repeat this investigation using a piezoelectric ceramic device rather than one containing a crystal.

2. Prepare a research report for your teacher describing how piezoelectric crystals are key elements in microphones. Read it to your classmates.

3. Prepare a research report on semiconductor diodes. Submit it to your teacher.

Bioelectricity: It's Shocking!

✔ INSTRUCTIONAL OBJECTIVES

Students will be able to

- explain the role of electricity in living organisms.
- draw conclusions based on observations.
- explain the differences between static electricity, current electricity, and biological electricity.

🌐 NATIONAL SCIENCE STANDARDS ADDRESSED

Students demonstrate a conceptual understanding of

- electrical motions or transformations of energy.
- big ideas and unifying concepts such as cause and effect.
- regulation and behavior, such as that of the senses, and response to environmental stimuli.

Students demonstrate scientific inquiry and problem-solving skills by

- identifying the problem and evaluating the outcomes of the investigation.
- working individually and in teams to collect and share information and ideas.

Students demonstrate effective scientific communication by

- arguing from evidence such as data collected through their own experimentation.
- explaining scientific concepts to other students.

✂ MATERIALS

- Small piece of aluminum foil
- Percussion hammer
- Pen flashlight
- *Optional*—EKG setup or video/laserdisc of EKG
- 🖐 *Optional*—electric catfish in 55-gallon tank or visit to tropical fish store/aquarium
- Human subject with metal fillings in dental cavities
- *Optional*—demonstration, laserdisc, or video of electrical stimulation of gastrocnemius muscle of a pithed frog
- Chart or diagram of knee-jerk reflex
- Stopwatch or clock

🖐 = Safety icon

HELPFUL HINTS AND DISCUSSION

Time frame: One class period, plus possible field trips

Structure: Individually and in cooperative learning groups

Location: In class, and on possible trip to aquarium or tropical fish store

In this activity, students will learn how current and static electricity differ from biological electricity. There are three optional activities and it is recommended that you have your students experience at least one.

You may wish to demonstrate a pithed frog to show the stimulation of the gastrocnemius muscle, or you may show this by means of a laserdisc or video cassette. Similarly, if your school does not have EKG equipment or an oscilloscope, you might want to demonstrate or show an EKG by means of an audiovisual device. Be sure to mention that an EKG is measuring the electricity flowing through the membranes of the heart muscle *before* the muscles contract. It does not measure the actual muscle movement.

The electric catfish *(Malapterus electricus)* is a rather large fish, about 60 cm in length, and it is carnivorous and aggressive. It is quite hardy and can be kept alone in a 50-gallon tank and fed raw meat or prepared food from a tropical fish store. These fish can produce up to 400 volts so precautions must be taken to insure that your students *don't, under any circumstances, touch* the fish. It would be in order to caution them about putting their hands in the water with an electric catfish. A public aquarium is perhaps the best place to learn about electric fish because its tanks are usually equipped with probes and amplifiers. Pet stores do not always have electric catfish in stock, so call before sending students.

Demonstrate the proper position of the knees and the location of the patella before your class does the knee-jerk reflex. You may wish to eliminate the use of the flashlight when they are doing the iris reflex. The flashlight should not have powerful light. A flashlight powered by 2 AAA batteries is sufficient.

ADAPTATIONS FOR HIGH AND LOW ACHIEVERS

High Achievers: Have these students carry out the Follow-up Activities. They should help the low achievers carry out the reflex procedures.

Low Achievers: Provide a glossary and reference material for boldfaced terms in this activity. They should be provided with detailed instructions regarding their tasks in this activity.

SCORING RUBRIC

Full credit should be given to those students whose observations and answers are complete and accurate. Students should answer the Concluding Questions in complete sentences. Give extra credit to those students who complete the Follow-up Activities.

INTERNET TIE-INS　http://www.epub.org.br/cm/n06/historia/bioelectr4_i.htm
http://www.kidshealth.org/kid/question/brain.html
http://abcnews.go.com/sections/science/dyehard/dye21.html

QUIZ　1. How does electricity in living things differ from ordinary home electricity?
2. Name three activities in your body that require electricity.

Bioelectricity: It's Shocking!

⚡ BEFORE YOU BEGIN ⚡

As you read this page, electrical messages are going to your brain by means of nerve cells called **neurons.** Actually, each and every living cell in our bodies has tiny electrical currents in it. Put all that electricity together and you have enough electricity to light a small bulb, such as the one in your refrigerator. You may already know that in current electricity, electrons flow from atom to atom in a conductor. However, in living things, electricity is produced by **ions.** Ions are either **atoms** or **molecules** that have either gained or lost electrons. Thus, ions can be positively or negatively charged. In our bodies, the heart muscles and all of our muscles use electricity when they are working, and hormones are **secreted** from **endocrine glands** by electricity. All cells have a tiny electrical voltage across their membranes. We have already mentioned that neurons use electricity. Electric fish use electricity for protection and to capture their food. More will be said about them later on.

🐟 MATERIALS

- Small piece of aluminum foil
- Percussion hammer
- Pen flashlight
- *Optional*—EKG setup or video/laserdisc of EKG
- 🖐 *Optional*—electric catfish in 55-gallon tank or visit to tropical fish store/aquarium
- Human subject with metal fillings in dental cavities

- *Optional*—demonstration, laserdisc, or video of electrical stimulation of gastrocnemius muscle of a pithed frog
- Chart or diagram of knee-jerk reflex
- Stopwatch or clock

🖐 = Safety icon

📦 PROCEDURE

1. For this first step, you will need someone who has metal fillings in his or her dental cavities. It could be yourself, a member of your cooperative learning team, an adult in school, or a family member at home. Fold the piece of metal foil several times so that it fits easily in the mouth and can be chewed.

2. Ask the person with the metal fillings to chew on the foil about six times. Ask the chewer to describe the sensation he notices and any change in taste he detects with his tongue. (The tongue is very sensitive to electricity.) Write the chewer's observations in the Data Collection and Analysis section.

3. In order to see an electrical pathway through the neurons, consult the chart or diagram of the knee-jerk reflex, which you will perform on one of your team members. Note that there are only three nerves involved. Have a team member sit with one knee crossed over the other.

(continued) 🔥

Bioelectricity: It's Shocking! *(continued)*

4. When the team member is relaxed, using the percussion hammer, lightly strike the upper knee just below the knee cap (**patella**). Observe the reaction. Indicate in the Data Collection and Analysis section how quickly the response occurred and if the subject had to think about reacting to the percussion hammer.

5. Another reflex involving muscle movement is the contracting of the **iris** of the eye due to light.

6. Seat one member of your team comfortably in a chair. Ask him or her to cover one eye with a hand for at least one minute. At the end of the minute, ask the subject to remove his or her hand. Note the size of the **pupil** (the black circle in the center of the eye). Shine the flashlight into that eye and note the change in the size of the pupil. The muscles around the pupil are in the **iris** of the eye. Write your observations in the Data Collection and Analysis section.

7. Observe the optional procedures your teacher may have selected for you, including the electrical stimulation of a frog muscle and an electrocardiogram of a person. Your teacher also may want you to visit a nearby aquarium or tropical fish store, or possibly to see an electric catfish in a large tank in your school. 🖐 **Caution: Do not touch these fish! They can produce an electric current of 400 volts.** The purpose of each optional activity is to familiarize you with another aspect of electrical energy in living things.

📏 DATA COLLECTION AND ANALYSIS

1. What sensations did the chewer notice?_____

_____

_____

2. What kind of taste did the chewer experience? _____

_____

3. How quickly did the subject react to the hammer blow? _____

_____

4. Did the subject decide to react or did the reaction occur without thought? _____

_____

_____

5. In the optional activity(ies) you witnessed, what role did electricity play? Describe in as much detail as possible._____

_____

_____

_____

_____

(continued)

Bioelectricity: It's Shocking! *(continued)*

CONCLUDING QUESTIONS

1. Compare the mouth of the person chewing the aluminum foil to a wet cell battery. What is the role of the aluminum foil, the metal fillings, the saliva, and the tongue? _____

   _____

   _____

   _____

   _____

   _____

2. There are a number of electric fish; however, there are no electric land animals. Why? (Hint: Refer to what you have learned about conductors and insulators.) _____

   _____

   _____

   _____

⚡ Follow-up Activities ⚡

1. Research the organs in electric fish that are responsible for producing electricity. Report your findings to your classmates.

2. Research various kinds of electrocardiograms, such as:
   Echocardiograms
   Stress electrocardiograms
   Present your findings to your teacher.

Plating: Is It Really Electrolysis?

✔ INSTRUCTIONAL OBJECTIVES

Students will be able to

- explain the process of electroplating.
- compare electroplating to electrolysis.
- draw conclusions based on observations.

🌐 NATIONAL SCIENCE STANDARDS ADDRESSED

Students demonstrate a conceptual understanding of

- electrical motions or transformations of energy.
- big ideas and unifying concepts such as cause and effect.

Students demonstrate scientific inquiry and problem-solving skills by

- identifying the problem and evaluating the outcomes of the investigation.
- working individually and in teams to collect and share information and ideas.
- recognizing, analyzing, and critiquing alternative explanations.

Students demonstrate effective scientific communication by

- arguing from evidence such as data collected through their own experimentation.
- explaining scientific concepts to other students.

Students demonstrate competence with the tools and techniques of science by

- using tools and technology to collect data.

✂ MATERIALS

- Three 500-ml beakers
- Lantern battery, 6 volts
- Two pieces of wire, 20 cm in length, with alligator clips on each end
- Piece of 12-gauge copper wire with insulation removed, 10 cm in length
- Two stainless steel spoons
- Coarse sandpaper
- Wooden spoon

- Copper sulfate solution (600 ml)
- Silver nitrate solution (300 ml)
- Silver foil
- Powdered graphite
- Watch or clock

HELPFUL HINTS AND DISCUSSION

Time frame: One class period
Structure: Individually and in cooperative learning groups
Location: In class

In this activity, your students will carry out three examples of electroplating. Silver can be obtained from most scientific supply companies. Gold foil for one of the Follow-up Activities is also available from scientific supply companies. The procedural steps are straightforward and should present no difficulty. Test the batteries to be sure that they have not lost their charges. Do not allow the students to use their electroplated spoons for eating.

ADAPTATIONS FOR HIGH AND LOW ACHIEVERS

High Achievers: Have these students carry out the Follow-up Activities. Have these students figure out the chemical equations involved in electroplating.

Low Achievers: Provide a glossary and reference material for boldfaced terms in this activity. These students should be encouraged to work with the high achievers.

SCORING RUBRIC

Full credit should be given to those students whose observations and answers are complete and accurate. Students should answer the Concluding Questions in complete sentences. Give extra credit to those students who complete the Follow-up Activities.

🖥 INTERNET TIE-INS

http://www.chemtutor.com/redox.htm
http://www.samson24k.com/museum.htm

❓ QUIZ

1. What is electroplating?
2. What kind of materials are the easiest to electroplate?

Name _____ Date _____

Electroplating: Is It Really Electrolysis?

⚡ BEFORE YOU BEGIN ⚡

Electroplating is used in making jewelry and flatware (knives, forks, and spoons). Usually a cheap metal is coated with a more expensive metal. This makes the object that has been plated more beautiful and also less likely to corrode. In order for any kind of electroplating to take place you need three things. You need **direct current**, which can be provided by a **dry cell battery**. You also need two **electrodes**. One electrode is the material that will be coated. The other is made of the coating material. Both electrodes must be able to conduct electricity. Metals conduct electricity well and so are easy to plate. Finally, you need an **electrolyte**. This is a solution that conducts electricity. In this activity, you will plate a stainless steel spoon with copper. Then, you will try to plate a **nonconductor** by coating it with graphite particles. The graphite, which is a form of carbon, is a good conductor. Hopefully it will make the surface of the nonconductor into a conductor of electricity.

The object to be plated is connected to the negative terminal of the DC source by a wire as shown in the diagram in the Procedure section. Another wire connects the positive terminal of the DC source to a bar of the plating metal. When you turn on the current, the electrons flow to the object to be plated and gather there. The object is now negatively charged. The positive metal **ions** from the plating metal in the electrolyte are then attracted to the negatively charged object. When the metal ions touch the object to be plated, they greedily grab the electrons and change from ions to metal **atoms**, which are neutral and have no charge. As more and more atoms accumulate on the object, the object becomes coated by them.

✂ MATERIALS

- Three 500-ml beakers
- Lantern battery, 6 volts
- Two pieces of wire, 20 cm in length, with alligator clips on each end
- Piece of 12-gauge copper wire with insulation removed, 10 cm in length
- Two stainless steel spoons

- Coarse sandpaper
- Wooden spoon
- Copper sulfate solution (600 ml)
- Silver nitrate solution (300 ml)
- Silver foil
- Powdered graphite
- Watch or clock

◆ PROCEDURE

1. Pour 300 ml of the copper sulfate solution into one of the 500-ml beakers.

2. Using your hands, twist the bare copper wire into a loose coil, as shown in Figure 1 at right.

Figure 1

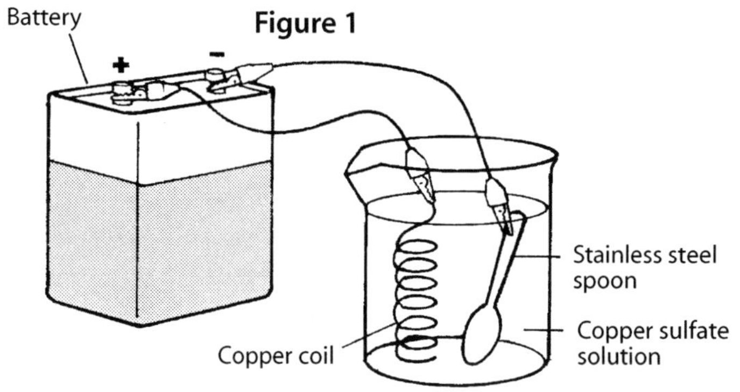

Battery

Copper coil — Stainless steel spoon — Copper sulfate solution

(continued)

Electroplating: Is It Really Electrolysis? *(continued)*

3. Connect one end of a wire with alligator clips to the positive terminal of the battery. Attach the other end to the copper wire coil.

4. Attach the other wire with alligator clips to a stainless steel spoon. Attach the other end to the negative terminal of the battery.

5. Insert both the copper wire coil and the stainless steel spoon into the copper sulfate solution.

6. Using the watch or clock, record how long it takes to see the first sign of plating in the table in the Data Collection and Analysis section.

7. Let the reaction continue until the spoon is completely plated with copper. Record the time it took in the table in the Data Collection and Analysis section.

8. Roughen as much of the surface of the wooden spoon as possible with the sandpaper. The rougher the surface, the better the graphite powder will stick to the spoon. Sprinkle the spoon with graphite, trying to get as much graphite as possible onto the spoon's surface.

9. Pour the remaining 300 ml of copper sulfate into another beaker.

10. Remove the copper-plated spoon from the first beaker and replace it with the graphite-coated spoon.

11. Place both the copper coil and the graphite-coated wooden spoon in the second beaker of copper sulfate.

12. Repeat Steps 6 and 7.

13. Pour the silver nitrate solution into the third beaker.

14. Connect the positive pole of the battery to the silver foil.

15. Connect the negative pole of the battery to the other stainless steel spoon.

16. Place both the silver foil and the stainless steel spoon in the silver nitrate solution.

17. Using the watch or clock, record how long it takes to see the first sign of plating in the table in the Data Collection and Analysis section.

18. Let the reaction continue until the spoon is completely plated with silver. Record the time it took in the table in the Data Collection and Analysis section.

(continued)

Electroplating: Is It Really Electrolysis? *(continued)*

DATA COLLECTION AND ANALYSIS

Solution	Object	Time—First Sign of Plating	Time for Complete Plating
Copper sulfate	Stainless steel spoon		
Copper sulfate	Wooden spoon		
Silver nitrate	Stainless steel spoon		

1. How does the copper plating on the wooden spoon compare to the plating on the stainless steel spoon? _____

   _____

2. How did the times for plating copper compare to the times for plating silver? _____

   _____

   _____

CONCLUDING QUESTIONS

1. What determines the amount of metal that will be deposited on an object? _____

   _____

   _____

2. What would happen if you reversed the wires? That is, if you connected the object to be plated to the positive pole and the plating metal to the negative pole?_____

   _____

   _____

3. Compare the process of electroplating to that of electrolysis. Research the process of electrolysis in a chemistry textbook. Save time by consulting the index. How are these processes alike?

   _____

   _____

   How do they differ?_____

   _____

⚡ Follow-up Activities ⚡

1. Repeat any of the three investigations using two or three lantern batteries. How do your times compare?
2. Prepare a research report for your teacher describing how tin cans are electroplated. Read it to your classmates.
3. Use gold foil to electroplate a metal fork.

J. WESTON

WALCH

PUBLISHER

Share Your Bright Ideas with Us!

We want to hear from you! Your valuable comments and suggestions will help us meet your current and future classroom needs.

Your name_____Date_____

School name_____Phone_____

School address_____

Grade level taught_____Subject area(s) taught_____Average class size_____

Where did you purchase this publication?_____

Was your salesperson knowledgeable about this product?　Yes_____　　No_____

What monies were used to purchase this product?

___School supplemental budget　　___Federal/state funding　　___Personal

Please "grade" this Walch publication according to the following criteria:

Quality of service you received when purchasing	A	B	C	D	F
Ease of use	A	B	C	D	F
Quality of content	A	B	C	D	F
Page layout	A	B	C	D	F
Organization of material	A	B	C	D	F
Suitability for grade level	A	B	C	D	F
Instructional value	A	B	C	D	F

COMMENTS:_____

_____

What specific supplemental materials would help you meet your current—or future—instructional needs?

_____

Have you used other Walch publications? If so, which ones?_____

May we use your comments in upcoming communications?　___Yes　___No

Please **FAX** this completed form to **207-772-3105,** or mail it to:

Product Development, J. Weston Walch, Publisher, P.O. Box 658, Portland, ME 04104-0658

We will send you a **FREE GIFT** as our way of thanking you for your feedback. **THANK YOU!**